# The Story of Humanity

## Ecology & Consequence



Ishi Nobu

The Story of Humanity: Ecology & Consequence

by Ishi Nobu

A *Spokes of the Wheel* Book

ISBN 978-1-948627-23-8


Web site: *www.ishinobu.com*
Email: *comment@ishinobu.com*



Produced in the United States of America.
Cover painting by Ishi Nobu.

1st edition, 2nd revision.



Other books by Ishi Nobu:

*The Red Pill: Mastering The Matrix*
*Clarity: The Path Inside*
*Unraveling Reality: Beyond the Veil of Existence*
*Spokes of the Wheel*:

1. *The Science of Existence*
2. *The Web of Life*
3. *The Elements of Evolution*
4. *The Ecology of Humans*
5. *The Echoes of the Mind*
6. *The Fruits of Civilization*
7. *The Pathos of Politics*
8. *The Hub of Being*

# Table of Contents

# Preface

Earth is an uncommonly lovely planet. Though subject to sporadic mass extinction events, Earth has hosted a lushness in life seldom seen. This watery, rocky orb about the Sun is truly a rare jewel.

The availability of freshwater regulates the diffusion and diversity of animals on land. Whereas moist times abet dispersal of terrestrial species, aridity sprouts spurts of speciation.

In one episode, an ape came down from the trees and pursued a global diaspora. We watched with fascination. This creature was unlike any other. It seemed an awfully odd experiment in obfuscated mind construction: setting the bar to self-realization too high.

A tragicomedy, humans. As quickly as they could, they wiped themselves out. The wry irony is that they considered themselves an epitome of intellect and reason.

The story of men is a tale of hubris. Education, wisdom, and comity should have been their abiding priorities, as they were when these apes were descending into humanity.

Instead, exploitation, profit, and power over others became the chief concerns of men. Compounding their error, men took liberty from women, their brightest jewel.

Our chronicle begins with the biology that defined the Earther experience. The plethora of ape races that converged to call themselves *human* adapted physically and mentally to the habitats they faced, and to each other.

The physical adaptations that crafted Earthers are remarkable. But they pale in comparison to the cerebral constructs which set the vector of their existence.

Human mental limitations were an incongruity to the physical power which Earthers garnered, via technologies, to alter their world. This disconnect is especially jarring considering the arrogance of men. What men considered their unique strengths were instead their worst weaknesses.

Like baboons, dominance hierarchies defined how human sociality worked. Civilizations emerged late and ossified into a primitive form.

Plutocracy ruled political economy throughout history. The collective discipline that would have ensured equity and prosperity for endless generations eluded Earthers.

Modernity added a chimeric gloss of justice in the guise of democracy. Popular elections provided a frustration vent but altered societal dynamics in only token gestures. Unknowing to the end, Earthers elected to kill themselves off.

Despite the temporal brevity of their existence, the archive of Earthers is extensive. They did love to show off.

What we have done in this brief volume is to distill the human chronicle down to a sketch of what transpired, with emphasis on the psychology – the 'why' – of what happened.

~ The Scribes of Sculptor
(Planet Signore of Beta Sculptoris, Galaxy NGC 253)

Earth

# A Fateful Ape

A dramatic shift in Earth's climate brought forth a fateful ape. A later climate drama ushered that ape's demise. The beginning owed to geology, the end to technology. Recorded human history spanned a mere 6,000 years.

A consistent system of dating helps chronicle the human experience. Earthers were extinct by what the natives would have reckoned to be the year 2100 CE (common era) in the dominant Gregorian calendar: gone by the dawn of the 22nd century CE.

The exact year of the fall is uncertain. Small communities and a few hardy souls may have survived for a few decades after the fall. Recorded history faltered 3 decades before the fall, as global industrial civilization crumbled.

For consistency, dates are given in years "before the fall" (BF); TBF for thousands of years BF, and MBF for millions of years BF.

This looking-back is an inverse accounting from traditional date-keeping, which progresses forward into an indefinite future. For instance, Earthers denoted the 1st (their last) century as the 21st century. The traditional year 2000 translates to 100 BF.

A reverse chronology creates a countdown. Sometime "early" in an era, century, or decade has a higher number than sometime "late," which happens at a lower date.

The calendrical use of before-the-fall aptly shows how close to extinction events occurred, and so sharpens the perspective on the vectors of consequence wrought by Earth's clothed apes.

❁❁❁

The Oligocene lasted from 34 to 23 MBF. That geological epoch began warm but cooled in the close.

There was a rapid radiation of animal species during the Oligocene, especially among mammals. Apes descended from monkeys near the end of that epoch.

The Miocene followed (23–5.3 MBF). Climatic swings intensified during the Miocene. Aridity rose.

## Earth

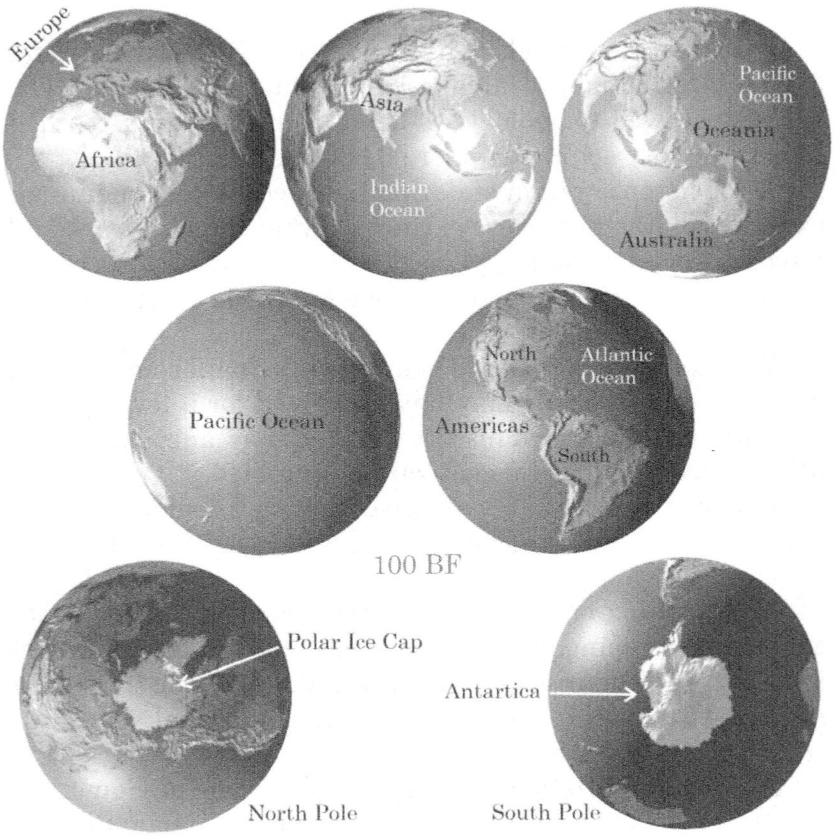


A drier world meant forests gave way to woodlands and savanna in many places. For arboreal animals, it was an invitation to be on the ground.

The solicitation was met by speciation among the apes in Africa. Some of those creatures that came down from the trees became *hominids*: more humanoid than anthropoid.

Naturally nomadic and randy, dispersal and interbreeding were common among hominids. There were dozens of hominid lineages. Many died out. The clade that carried on were *hominins*: the evolutionary branch of apes that begat humans (Earthers).

The basic body plan of hominids was established 18 million years before the fall. The multitude of species that arose were variations on that theme.

¤✧¤

A sharp drop in global temperature 115 thousand years before the fall was the start of sporadic ice ages. The last glacial spell Earthers experienced ended only 12.5 TBF.

Earth is on the cool side of the habitable zone provided by the Sun. Icehouse has been more frequent than hothouse.

It was not temperature but its dramatic flux that spurred evolution and altered life's prospects on Earth, especially for land animals. Mass extinctions result when climatic shifts rapidly ratchet beyond what can be adaptively tolerated. Such an event spelt the end of the apes who technologically self-destructed.

Human self-annihilation was not the first time that organisms created a mass extinction event. 2.45 billion years before the fall, marine cyanobacteria infused the sea surface and atmosphere with oxygen, slaughtering anaerobes who could not tolerate $O_2$. In part, human self-extinction also involved polluting the atmosphere.

Animals arose 890 million years before the fall. The first animals to drive their own demise were sea worms, 542 MBF. These persistent wigglers ravaged marine microbial populations and then succumbed to their exuberance. Despite the doom, the upshot had an upside. Life's recovery saw a profound proliferation of new species.

Dinosaurs emerged when the world was warm. This was roughly 242 MBF. The balmy climate owed partly to the configuration of continents. Much of Earth's landmass was concentrated into the supercontinent Pangea.

Pangea's partitioning resulted in isolated populations. Distinct biomes emerged, engendering speciation among all sorts of life. While some lineages of dinosaurs supersized to be fearsome beasts, one diminished to become the exquisite creatures called *birds.*

The apes that called themselves human began their identifiable journey 8 MBF, as grasslands expanded and the Himalayas began their ascent.

A critical factor driving plant evolution was the availability of water. This was apparent in root system development, and in the ascension of flowering plants (angiosperms). Compared to their predecessors, angiosperms concocted numerous innovations to better manage both water and herbivory.

The angiosperm most astonishingly adapted to aridity was grass. The grains that would become integral to the human diet were the seeds of grasses.

Aridity also shaped hominids. In 2 notable instances, cycling from a wetter climate to a drier one spurred revision in the hominid family tree, as well as a decided dip in numbers.

The 1st instance occurred ~2.7 MBF. For over 3 million years, the dominant hominids had been in the genus *Australopithecus*. They gave way to *Homo* when the climate dried.

Australopiths were short, stocky, and hairy. In appearance, their direct descendant was *Paranthropus*, who was a quite robust ape. Despite its tough build, paranthrops died out, unable to handle wrenching climate changes.

That left *Homo*, the genus from which humanity sprang. With its taller, slimmer build, *Homo* was more wiry than other apes. More surface-area-to-volume improved heat dissipation. The effect was better body water conservation.

Other adaptations accompanied greater gracility. From 7 MBF, hominids had become more mobile by walking upright. Walking got even easier as this creature evolved. The stride of *Homo* refined from its ancestors.

Coupled with stamina, hominid bipedalism broadened foraging range. This out of necessity. The high-quality plant foods once abundant in dense forests were more dispersed over woodlands and fields.

As forests gave way to grasslands, hominids lost body hair and gained sweat glands attuned to cooling. These adaptations improved shedding heat and keeping a safe body temperature while on the move.

The enhanced ability to sweat had a downside: it elevated the risk of dehydration. That vulnerability made hominids more reliant on freshwater than their primate cousins, and far more so than desert-adapted animals such as sheep, goats, and camels. A goat may shed over 1/3rd of its body water before expiring from the loss. A hominid was at risk of death by dehydration with more than a 10% water loss.

Hominid dietary flexibility was a defense against dehydration. In dry biomes, these creatures sought foods laden with water, such as succulent fruits.

While Earthers evolved to use less water despite more profuse sweating, they still needed to drink water. That need drove hominid migrations and set where these creatures settled.

The 2nd instance where a spike in aridity prompted hominid revision happened 1.75 MBF. *Homo erectus* arose, along with several similar subspecies. Its taller, lither skeleton was nearly indistinguishable from modern humans.

*H. erectus* was the first hominid to leave Africa. This ape's confidence showed through its technology. The first double-sided hand axes were made by *H. erectus.*

Global climatic patterns are influenced by Earth's solar situation, which occurs in extended cycles. Earth's orbit has a 95,000-year cycle. Earth's tilt cycles over 42,000 years, and its distance from the Sun every 21,000 years.

Though aridity spurred speciation, wetter climate abetted dispersal, as freshwater was more readily available. The hominid diaspora went intercontinental as this wanderer trekked throughout Asia, Europe, and beyond: taking to rafts and boats to traverse the seas.

Though they mostly ate plant foodstuffs, hominids were omnivores. Meat was harder to come by, and raw meat carried the risk of disease from microbes. Yet hominid yen for meat was strong, despite it being less digestible. Meat became more of a staple than with other apes.

All sorts of animals, from insects on up, were feasted upon when they could be caught or killed. Then there was another meaty delicacy close at hand.

Cannibalism was widespread. "Cannibalism was, at one time, practically universal; it has been found in nearly all primitive peoples," wrote 2nd-century-BF historian Will Durant.

Hominid taste for meat had an unintended consequence: zoonotic diseases. Pathogens used to other animals developed their own taste for these furless apes.

Earthers eating other animals provoked period plagues. Epidemics repeatedly decimated human populations.

Hunting was a hard-earned skill, demanding both tactics and technology. Men took to hunting as a sporting challenge.

Meanwhile, women were key to survival. Women were the repository of dietary knowledge: which plants, and which parts of them, were nutritious. The acumen essential to the eventuation of agriculture owed to women.

Given their nurturing nature, women were the ones who first mastered medicine: learning the magic of herbs and turning that sagacity into lore that was passed through the generations. Until their demise, Earther's primary medicines were from plants.

During Earther's salad days, women wise in the ways of Nature were especially esteemed. Among them arose enlightened shamans and the cultivation of the earliest spiritual practices.

Though agriculture was to irrevocably alter hominin sociality, conjure economics, and have grievous planetary consequence, cultivation was nothing novel. Amoeba bred bacteria to eat. Ants tended aphids. Termites farmed fungus.

¤✧¤

Hominids diverged from other apes owing to *neoteny*: retention of traits in adults previously seen only in the young. Chordates – the phylum of vertebrates – arose over a half-million years BF via neoteny. This adaptive technique also produced flightless insects, salamanders, and birds.

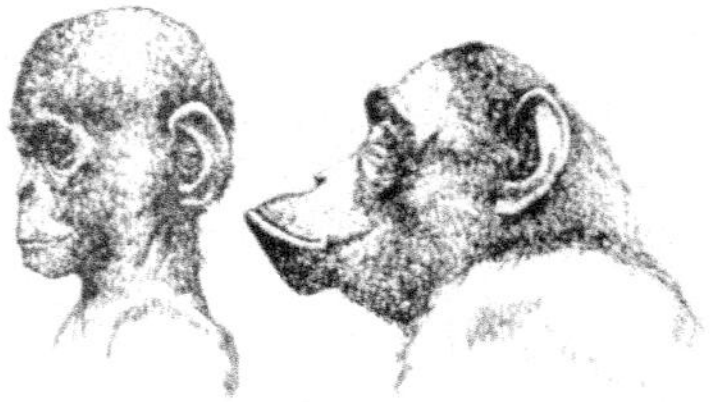

Juvenile *Chimp* Adult

Neoteny shifted hominid descent onto a distinct track in numerous ways. It produced a larger ratio of brain to body size, lighter bones, a flattened and broader face, larger eyes, smaller nose, smaller teeth and jaw, and glabrousness (hairlessness).

Human limbs and body posture were indicative of neoteny. Limbs were proportionately short compared to torso length, longer leg than arm length, a juvenile foot structure, and upright stance.

Many styles of bipedal locomotion are possible. With their pelvis and leg structure, many birds are as apt to hop as walk

– though not all. The fastest animal on 2 legs when Earthers were extant was the range-dwelling ostrich, the biggest flightless bird.

Other apes took longer strides by rotating their hips. Hominids evolved to longer legs and minimal hip swivel when they walked.

Though other apes might walk on 2 legs, an assist with the arms – knuckle walking – was used to gallop on the ground. Hominids ran on their feet alone. Though taking shorter strides, swinging arms optimized efficiency.

For Earthers, walking was excellent exercise, both physically and mentally. Walking elicited a holistic harmony of balance that attuned attention, thus naturally facilitating heightened awareness.

Humans evolved to need considerable physical activity – especially walking – to stay fit. By contrast, other apes may spend their days lounging about and retain their health.

Neoteny typically results in larger individuals with longer lifespans. It did so with humans.

Neoteny also begat birds. They were a notable exception in size and longevity. To gain flight, an essential facet of avian evolution was downsizing. Small size has its price. Most birds live only 3 decades or less.

One of the hallmarks of mammals is their fur. A thick coat of hair provides protection as well as abetting tactility. For mammals, glabrousness – lack of hair – was often associated with neoteny. To facilitate heat dissipation via sweating, hominids went furless.

To accommodate their subterranean lifestyle, the sand puppy of east Africa also lost its fur, albeit retaining scattered tactile hairs over its body.

Becoming glabrous was one of several neoteny traits for the sand puppy. Another was eusociality, which is the epitome of cooperative sociality. The fate of Earthers would have been different had they the inborn comity of sand puppies.

Neoteny in hominid genitals was seen by lack of a penis bone (baculum) in males, and, in females, a forward-facing vagina and the presence of a hymen. These physical changes altered the emotional context of sex for these creatures: arousing an intimacy in the act.

Neoteny affected perception of bodily beauty, especially in males. Female faces that were more juvenilized were most appealing to men. Less neoteny was unattractive, regardless of age. As women chose a mate largely on concerns other than physicality, neoteny was not nearly as looming an issue in male allure.

Aesthetics were important to hominins. While beauty was always a subjective appreciation, behind it was biology.

What the senses let in is a limited range. What the mind casts as appealing depends upon its proclivity in patterning. Both are biological processes.

The female was the epitome of human aesthetics. Beauty there had 2 facets: the face and overall body shape.

The face of a strikingly beautiful woman had some combination of larger-than-average eyes, higher-than-average forehead, fuller-than-average lips, shorter-than-average jaw, and smaller-than-average chin and nose. Unblemished skin, high cheekbones, and lustrous hair completed the picture of beauty. All of these are features of physical health and are indicative of fertile mating potential.

Symmetrical faces were more attractive than unsymmetrical ones. Men with high symmetry typically had more sexual partners and induced more copulatory orgasms.

Bodily symmetry was an honest signal of health. All animals consider symmetry appealing, rightly taking it as a cue to the ability to produce robust offspring.

The admired hourglass figure of a woman was an evolutionary refinement. "The sex appeal of rounded female buttocks and plump breasts is both universal and unique to the human primate," observed 1st-century-BF primatologist Carole Jahme.

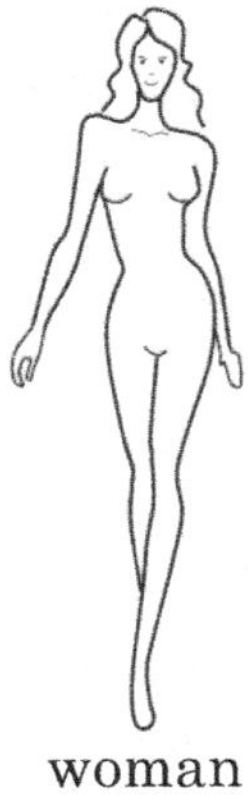

woman

In its origins, clothing was a form of ornament, not a cloak against the cold. Regardless of epoch, women asked of clothing not to cover their nakedness, but to suggest or enhance their physical allure.

The teats of other female apes swelled only to store milk. The abiding plump bosom of women was an anomaly among primates. Only human females developed a pronounced bust before menarche and retained it post-menopause. As such, breasts did not indicate fertility.

Nor was a woman's bosom indicative of lactation. Bigger breasts did not necessarily make more milk.

The symmetry of bosoms suggested fitness. Fluctuating breast asymmetry was higher in women with larger teats, and in women without children.

Unlike men, women first stored fat in their buttocks, not around the abdomen. Hence, the waist of a healthy woman was slimmer than her hips. Other female primates did not deposit fat on the rump. Female gorillas were exemplary in keeping a skinny bum while putting on abdominal girth, as hominid males did.

Female breasts had a straightforward appeal to males. For the derriere, there was an angle to the appreciation. The admiration of an hourglass figure on the back side owed to a geometric inclination.

A 45.5° incline from rump to spine was most attractive to men. Not coincidentally, it also provided the ideal angle to let pregnant women balance their weight over their hips. Further, this angle was beneficial in giving birth, which was naturally done standing up. Standing facilitated dilation, blood flow, push, and the descent and rotation of a birthing baby.

Oddly, late-century birthing was done with the woman lying down. Such was exemplary of the cognitive decline of Earthers in the centuries preceding their demise.

Late-century humans abandoned many sensible health practices. For example, through lack of exercise and chronic overeating, most 1st-century-BF moderns were fat, many to grotesquerie. Their lives were debilitated and shortened as a result.

⌑✧⌑

Animal species exhibit distinct precocial and altricial strategies which define development. The ability of parents to provide care corresponds with the selected strategy.

The mothers of precocial birds lay protein-rich eggs, thereby letting their young hatch in the fledgling stage. Such chicks can protect themselves from birth. These birds have relatively little post-natal involvement with their offspring.

Altricial bird eggs contribute fewer nutrients in the prenatal stage. Eggs are smaller. Hatchlings need attention and protection.

Altricial avian young typically grow faster than the precocial sort. This quick development shortens the nestling stage, when predation risk is highest.

In the instance of mammals, large adult body size favors production of big, precocious young which develop via a longer gestation period. Large young are associated with small litter size, an extended reproductive period, and migratory behavior. This strategy was apparent in many ungulates (hoofed creatures), which lived on the move. It also applied to hominids, who similarly foraged over large ranges.

During their descent, hominids became increasingly altricial, albeit with a unique combination of precocial and altricial development. In terms of potential lifespan, physical development was relatively quick. More prominently altricial mammals, including many rodents, remain largely immobile and undeveloped until grown to near the stature of their parents.

In contrast to their precocial physicality, humans were very altricial in their mental faculties. Youth were slow to gain acumen. This coincided with their gregarious nature. Prolonged childhood allowed more time for bonding among peers, as well as education in the practical arts and folkways.

Sociality was not a cultural acquisition. It was instead sewn into the hominin psyche. Being an altricial animal meant that sociability was innate.

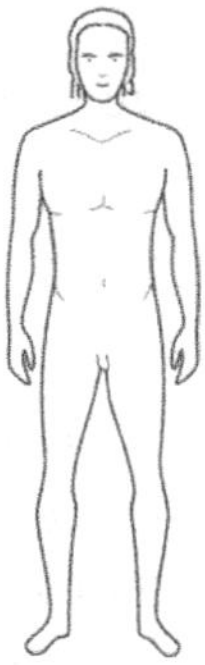
man

The need to belong was strong. Family formed the social nexus. Its bonds were the core from which clans formed, and through which cultures grew.

As with other apes, males suffered a pronounced dissonance in their sociality. Men had an inclination for independent action that ran counter to the harmonious productivity that came from cooperative endeavor.

The continuing psychosocial contention of individualism versus communalism was a key aspect of how human groups evolved. This schism sharpened in the last age, when environmentally destructive technology heightened the importance of cooperation. (The 'last' age is what people in post-industrial societies called "modern" times.)

Like other apes, hominids were typically exogamous (outbreeding) and patrilocal (a woman joining a man's extended family). A few prehistoric groups were matrilocal, with women as the social nexus. Such clans were extant in central China, southern India, Oceania, and in the Amazon basin.

Besides fostering genetic diversity, exogamy encouraged intergroup exchange. Exogamy also mitigated conflict by establishing bonds between groups.

Historically, disputes over resources or territory were smoothed over through the exchange of nubile females. At the top rungs of the sociopolitical ladder, the practice of affinity via strategic marriage often secured peace among conflicting tribes. Such alliances through marriage bonds were common among the European upper social stratum from Roman times, 2 millennia BF, into the last age.

Among primates, hominids were unique in females having *menopause*: living well beyond their reproductive years. Only females had menopause.

A few other mammals had menopause. All were social and altricial. Orca and beluga whales had menopause.

The evolutionary intent of menopause was extra assistance in rearing offspring. Menopause facilitated women being grandmothers, and so help raise children, especially sons.

Male children were less hardy than females, and naturally foolhardier. With their independent spirits, boys were in greater need of social education than girls.

Exogamy and menopause intertwined in humans. Owing to exogamy, women gained greater social status the longer they were in a clan.

Without menopause, older women might potentially compete with younger ones for mating partners. Menopause eliminated that potential social friction.

Elephants were a contrast to Earthers. Elephant societies were segregated by sex. Elephants had male exogamy.

For elephants, menopause could not confer greater social status, nor did friction among females for mating exist. Hence, elephants did not undergo menopause.

As with other organisms, viruses propelled hominid evolution. ~30% of adaptive alterations in human proteins came as viral contributions. The human inclination toward addiction for enhanced pleasure stimulation was a viral implant.

Hominids lost physical strength in their descent. Flexibility in muscle mass afforded surviving on spotty food supplies. Many rodents also have this adaptation.

Stamina took precedence over strength and speed. This adaptation was both physical and mental.

Humans were capable of unnatural will. As with other beneficial traits, mental fortitude lessened in late-century Earthers, when mental illness became the norm.

Mental adaptation corresponded with bodily changes. The hominid mind refined in the eon of evolution that culminated in humanity.

A more vivid imagination compensated somewhat for physical diminishment. This was apparent in the progress of hominid technology.

Another association with neoteny was altered sociality. Animal domestication repeatedly resulted in neoteny, as dogs illustrated. Hominid neoteny also sponsored sociability.

Greater gullibility in hominids fostered group cohesion and somewhat lessened their aggression. People were generally bad at detecting deception. Most were readily led by comforting beliefs. This was apparent by how religions of fantasy became so ubiquitous among Earthers.

Human gullibility had a decided drawback, which was shown by contrasting people to dogs in their treatment of lying.

In a revelatory experiment, dogs were taught to follow the advice of an unknown person in choosing which of 2 bowls contained a hidden treat. These dogs then watched another unknown human move the treat from one bowl to the other while a 2nd unknown person watched. In other instances, the 2nd person was not present for the switch.

After this, the dogs ignored human advice if the person had not been present when the bowls were switched. They knew the person did not know which bowl had the treat.

More significantly, dogs knew when they were being lied to and acted accordingly. They ignored those who deceived.

People were much more likely than dogs to follow the advice of a known liar: children especially so, but even adults. Voting trends in democracies amply document this bizarre inclination.

Gullibility proved a severe weakness. People readily succumbed to deception and were susceptible to manipulation. Natural optimism and deferring to authorities who misled them retarded collective response to the dynamics which ushered self-extinction.

A key capability enhancing hominid survival was improved dexterity, courtesy of a hand precisely proportioned with an opposable thumb. This innovation arose more than 3 MBF, which – by no coincidence – is when the Stone Age began. Manual dexterity improved with later hominids, affording technological advance as far as their minds were able to creatively find solutions.

Human optimism emanated from their belief in the power of imagination, which people treasured. Albert Einstein, the most renowned modern physicist, once strangely said, "Imagination is more important than knowledge." The combination of cherished imagination and greasy gullibility proved to be a damning detriment to humanity.

Early hominids lost the sharp canine teeth of their ape ancestors, taking the bite out of biting an opponent as an armament. There was compensation at hand.

Around 2 MBF, numerous traits evolved which endowed greater facility for fighting. Among them were stronger wrists and a hand that could form a fist. This turned a delicate musculoskeletal system into a club.

Contemporaneously, the faces of males diverged from those of females. The facial bones that differed most were those in men that strengthened protection in the face from injury during fist fights.

The function of the fist as a weapon was reinforced by bipedality. Unlike other primates, apes and hominids walk on their heels. This body posture lends extra punching power. It also suggests that physical conflict was instrumental in hominin descent.

The evolutionary development for physical fighting shows that hominid gregariousness was leavened with viciousness. This seeming maladaptation signified the social struggle that haunted men throughout their existence: overcoming selfishness.

Upper body strength declined in later hominins, taking the punch out of fisticuffs as a decisive advantage. Rhetoric was the ready substitute, though that was often not as persuasive as it should have been. Many men still found brawling and war preferable to argument and negotiation: a trait which was never bred out, as the legacy of countless conflicts throughout man's history amply testified.

⌑✧⌑

Many animals use tools. But hominids seemed especially fond of them. Comparatively, other primates used tools sparingly. A host of adaptations – improved dexterity, loss of muscle mass, enlivened imagination, inclination to craftwork – enhanced hominid ability to survive via tools and weapons.

Hominids made wood tools for many millions of years. ~3.4 MBF they took to shaping stone: flint knapping. Thus marked the onset of the Stone Age.

Working stone was not easy. The shaping of rock tools by hand required considerable skill. Over time, the work progressed from scraping suitable stones to crafting well-hewn spearpoints. Several hominid species practiced this craft. Cultural interchange was likely.

Hominids mastered fire by 1.9 MBF. This crucial step came after extended observation and experimentation. Other animals also used fire, as did some plants. Certain seeds sensed fire as a way to figure that ground cover may be minimal, and so be the best time to sprout.

Controlled heating enhanced survivability against the elements (and large predators), and heightened health by unlocking the nutrients trapped within plant cells: a feat which chewing alone could not. Fire also extinguished the risk of microbial infection from eating meat, though not the health hazard of its overconsumption.

Much later, fire afforded the forging of metals into otherwise unattainable shapes. This was a late innovation. Metalworking only emerged 11 TBF. It independently developed around the world.

It is no coincidence that metalworking arose shortly after agriculture became the global modus operandi of survival. Hard soil left wood tools wanting.

¤✧¤

Earther technology often showed cleverness in its creation but environmental disregard in its application. In the end, humanity committed mass suicide in thrall to terrible technologies they knew to be destructive.

Being smart is a skill in creating outcomes which foster survival. The earliest technologies did just that. Implements made digging up roots easier and mashing them possible, rendering foods more nutritious. The ability to cook food took nutrition to a new level, enabling health to a degree unimaginable before.

Mastery of fire exemplified the nature of technology: the ability to envision possibilities for manipulating the environment to achieve a productive facility. The genesis of technology is the mind, not the hand. Dexterity is only as good as the mental acuity behind it.

The artifacts hominids created were statements of mental condition. Weaponry most dramatically exemplified this. An increasing ability to commit mass murder did nothing to give men pause as to what they were doing to the comity essential to sustain societies beyond relying upon brute force.

Genuine social progress would have culminated in the banning of technologies rather than their continued deployment. Few Earthers made this point, and very few technologies were banned. The Collective were wooed by technologies which were obviously bound to have deleterious effects along with their convenience.

To grasp significance, technology must be examined in its social and environmental contexts. The impacts of technology expanded with their advance: a compounding of consequence in every sense.

Earthers' self-destructive ways owed to cultural values which reflected their belief systems. We turn next to examine the human mind and the cultures which their minds imbued.

# The Reign of Thought

Thought is fraught. Reason is merely the creation of self-satisfaction by stringing together beads of logic. Belief is consolation via a facile sense of surety.

Theory and worry emanate from the same source: the imagination. Their only distinction is practicability.

Eschewing skepticism, Earthers indulged their thoughts: finding in them inspiration, grief, and comfort. Beliefs lubricated social interactions, creating tribes via shared visions and values. "Ideas shape the course of history," remarked 2nd-century-BF economist John Maynard Keynes.

The glue of belief was also their undoing. Refusal to reconsider traditional ideologies terminated the species. The mob rule of democracy doomed humanity.

The reign of thought began long before written language chronicled men's musings. Ceremonial burials took place a million years before settlements.

❀❀❀

There are several evolutionary strategies for sustenance. The self-reliant make their own food from basic chemical substrates. The earliest life necessarily did so. Some bacteria still do, by sipping hydrogen or mining their food from minerals.

Plants are Earth's most sophisticated autotroph. Leaves feed via photosynthesis. Meanwhile, roots secure mineral supplements.

All other life on Earth survives off the autotrophs. For heterotrophs, the skills of foraging, scavenging, and hunting are essential to staying alive.

There is a more vicious variant of heterotrophy. Roughly half of Earth's life forms adopt a ruthless, cunning strategy: they steal their food supply. These are parasites.

Parasitism is tricky. A host must be found and its defenses overcome.

The incursion is best done by not killing the prize, at least right away. The savviest parasites afford their host a natural lifespan.

Viruses were first to commit to parasitism. While keeping their wits about them, viruses disposed of the physical baggage needed for reproduction. This economy was critical in the viral playbook: stay tiny so as to slip into cells with the littlest fuss possible.

¤✧¤

Evolution is adaptation, aimed at survival. Adaptation emanates from the ecology of living. Innumerable instances of biological design evidence this.

Land on Earth is habitable by animals because of plants. Underlying the power of plants is photosynthesis, which transmutes molecules into food, using sunlight as a power source. Its incredible efficiency comes from astutely tapping into quantum mechanics at key junctures. Knowledge of all facets of existence is expressed in evolution.

Nature does not select. It proposes.

There are many flamboyant examples of creative adaptation. The saltation of turtles, with their ribs suddenly sprouting whole-body armor, is exemplary. The gears that juvenile planthoppers had on their legs, which helped them make accurate leaps, is another astonishment. Grasses' extraordinary loss tolerance (to herbivory) is a florid floral example.

Evolution is holistic, intertwining physicality and psychology into biomechanics. What organisms experience and learn shape their evolution.

From an evolutionary perspective, all organisms face strategic survival trade-offs. These complementary compromises conceptually comprise *life-history variables*. These variables range across the traits of an organism and correlate to each other. For instance, the size of a body matters for its metabolism, growth, survival risks, and advantages within its habitat. Smaller plants and animals may adapt quicker because they have faster rates of molecular evolution.

Life-history variables coincide with the habitat in which an organism lives. Ecology drives adaptation.

The key survival aid of hominids was their minds. Their destiny hinged on how they employed this instrument.

The conflation of attributing 'native intelligence' obscures the facets of memory, analysis, and problem-solving which comprise the intellect. This blanket label may nonetheless be insightful in comparisons.

The natural savvy of a life form is a life-history variable. The *manipularity-intelligence theory* relates innate shrewdness with manipularity: the ease by which an organism can manipulate its environment. Life with lesser ability to alter its habitat needs to be smarter to survive.

Possessing the least physicality, viruses are a most intelligent life form. Viruses possess the remarkable ability to self-evolve. They do so by selecting specific stratagems. These schemes to infect and replicate are physically expressed as molecular structures.

The only facility viruses have to change their habitat is to alter the functioning of those they infect. To ease their invasions, viruses unified life under the molecular regime of DNA. Though a plague, viruses have also been essential to the evolution of other organisms.

In manipulating host genetics, viruses passed many of their creative innovations to their hosts. 8% of human DNA derived from viruses, including key mechanics for development.

Like viruses, microbes individually have little ability to alter their environment. They too must rely upon their wiles. Viruses and microbes both treat molecules as information bundles, as well as appreciating whatever their physicality may offer.

Microbes use sophisticated quorum sensing to assess their social situation. Pluricellularity has a power to craft a fate which individual cells may only dream of.

Sessile plants survive by their wits. The head start of autotrophy – not needing to find food – belies the difficulties of living in a largely uncontrolled environment.

For plants, a universe of decisions must constantly be made for apt resource allocations. There are trade-offs between growth and defense. Uncertainty defines both weather and the probability of attack by savage herbivores.

The unsurpassed intelligence of plants is amply illustrated by how they conduct their sociality with other life. Though generous with their fruits, plants also have a sense of fairness with those they contract services from. Among the many bacteria and fungi that plants employ, freeloading is not tolerated.

Intelligence originates with perspicacity, which gives rise to comprehension. At the base of intelligence is awareness, along with the mental ability to manipulate symbols. Living is a constant challenge of awareness and concept management.

Problem-solving is only possible through counterfactual thought: imagining what might be. This essential skill of symbolic manipulation is innate to all life, albeit varying in its attributes and capacity. Holistic intellectual capacity in animals is indicated by how frequently they resort to trial and error to solve problems.

Limited to beaks for hands, birds commonly act only after they have mentally figured the steps for solving a complex task. Changing tack only happens when new information is revealed during implementation.

Like birds, dolphins could not easily manipulate their environment. Their restitution for limited manipularity was a generous aptitude for problem solving, abetted by a convivial sociality which engendered cooperative effort.

By contrast, hominids had unrivaled ability to control their environment. They constantly resorted to trial and error to solve problems. Human memory, and ability to mentally map out solutions, were feeble compared to rodents, which had much less facility for physical manipulation.

With their precision grip and adroit motility, humans simply did not need to be very smart. What they got in compensation for their limited savvy was a fertile imagination.

Sociality supplies the context for language evolution. Language is universal. Even viruses exchange information.

Language is necessarily symbolic. A honeybee dances directions to her hive mates of a discovered food supply, with movement and timing signifying. Cleaner fish symbolically advertise their services to foreign fish, which reply with their own postures.

Many animals craft complex sounds. Avian vocalizations are an epitome of sonic beauty.

All apes expressed themselves by facial expressions, bodily poses and gestures, and vocalizations. With more flexible bodies, dexterous hands, and versatile vocal cords, hominids evolved the means to more expansively express themselves.

Human language developed alongside adaptive refinements in symbolic manipulation and its articulation, culminating in music and writing. Yet these pale comparatively.

Based upon the qualities of manipulated molecules, the languages of plants and microbes exceeded those of humans in sophistication. Though they documented biochemistry and genetics, Earthers failed to appreciate the intelligence behind the production of these molecular artifacts. In their ignorance, men believed in their own superiority.

Olfaction works by recognizing the signature of odorants: vibrational attributes, not molecular structure.

Microbes and plants rely heavily on specific molecules to sustain themselves. While plants manufacture much of their food, certain minerals must be obtained by root mining or trade with fungi and bacteria. The language plants use to communicate their wants pivots on expressing subtle olfactory qualities about specific chemical elements: an elaborate nomenclature indeed.

Excepting some outlier hunter-gatherer groups, people generally struggled to name specific scents. This did not owe to a poor sense of smell. Earthers were able to sift through billions of distinct odors. They used scent to detect fear, stress, and sickness, and to assist them in selecting mates. What they lacked was language to express their olfaction. The reason: scents were intangible, and therefore abstract.

Human language evolved as a cultural reinforcement, cementing socially accepted belief systems. Language was also a window into how the human mind worked, and its limitations.

People's minds portrayed a world of objects. All human languages were object-oriented. Nouns outnumbered verbs in all vocabularies. Infants learned new nouns more rapidly and more easily than new verbs. Earther sentences were of things acting or being acted upon.

Human memory favored objects over processes. Memories subject to conscious recall were *declarative* (aka *explicit*, in being readily explainable). The declarative memory categories are episodic, semantic, and topological. Episodic (flashbulb) memories are of events. Semantic memories are of facts and concepts. Topographic memories are mental maps.

The contrast was *procedural* (*implicit*) memory: memories of skills. Procedural memories are of processes.

Procedural memories are essential to living. Yet humans struggled to explain them. This difficulty owed to obfuscation by their mind.

All is process. What appear as objects are localized gyres, with their ecologies running in perceptual slow motion. Yet the human mind deceitfully insisted on the naïve idea of inert solidity: objects.

There was a verb for volitional conceptualization: think. The broader verb *mentate* would be both natural and proper. But that word was never coined. Instead, mental activity was ubiquitously referred to by a noun: the *mind*, a nonexistent entity. For explanatory convenience only, this chronicle adheres to the 'mind' misnomer in describing mentation.

❀❀❀

Faculty in math is innate in all organisms. Even microbes must be able to tell when food supplies are plentiful or running short. They do so based upon estimated quantity.

Many fish live in schools, which probabilistically gives them better odds of escaping predation. Fish can tell at a glance a small difference of group sizes: a speedy comparative power beyond human ken.

Language aptitude is related to intelligence in the broadest sense. Hence, math and literacy were the 2 facets of tests moderns used to measure intelligence. Mathematics is its own language.

Arithmetic was the first and most basic branch of math to emerge in the earliest societies. The more primitive the society, the closer representation of its number system tended to sets of straight lines. The earliest writings of the Mesopotamians and Egyptians, from 5500 BF, showed vertical straight lines to signify quantities.

In the 25th century BF, Hindus in India made an advance in numeral designation. A straight line represented the number 1, but distinct symbols were used for greater quantities.

Meanwhile, the Etruscan civilization in Italy was still using tally sticks. These evolved into Roman numerals, with a notch for 5 (V) and a crosscut for 10 (X).

Roman mathematics were focused on commerce. They did not evolve beyond this basic purpose.

Roman numerals were similar to the Babylonian system, which first appeared 4200 BF. The Babylonians had the 1st known positional numeral system, in which the value of a particular digit depended upon the digit itself and its position within a larger number.

The Babylonians understood the notion of nothingness, but it was seen as a lack of a number, not a number unto itself. The Babylonians used a space to mark the nonexistence of a digit at a certain place.

In calculating Jupiter's orbit, Babylonian astronomers came close to discovering calculus. Their mathematical techniques in 2450 BF were long thought by historians to have developed only in the 8th century BF. "The Babylonians developed abstract mathematical, geometrical ideas about the connection between motion, position, and time that are so common to any modern physicist or mathematician," wrote 2nd-century-BF archeologist Mathieu Ossendrijver.

Arabs learned of the Hindu numbering system a millennium after its invention, in the 14th century BF. The Arabic system first appeared in European arithmetic in 1124 BF, using the 9 digits: 1, 2, 3, 4, 5, 6, 7, 8, 9.

Though a tremendous step forward, the lack of a digit for zero precluded critical arithmetic operations. Subtraction was problematic if not impossible.

Zero was slow to be incorporated into mathematics, partly from the primal fear of the void it represented. Most ancient peoples believed only nothingness and chaos were extant before the universe came to be. Mathematically echoing this belief, ancient Hebrews, Greeks, and Romans had a revulsion to zero. Like mastering fire in the physical world, accepting zero propelled human mathematics to a new level.

Another number of immense significance is $\sqrt{-1}$. The number repeatedly cropped up, and was shunted aside, for a millennium.

The ancient Egyptians knew how to roughly calculate the volume (frustum) of a pyramid: quite an accomplishment for a people who had no knowledge of integral calculus. In figuring a frustum, the ancient Egyptians encountered $\sqrt{-1}$. They ignored the result and its implications. So did other ancient civilizations.

Negative attitudes about negative numbers were longstanding. The ominous sense was that there was something seriously wrong about a tally that not only could not exist but is also an inherent deficit.

Negative numbers were long thought an inexplicable pit of quantity. As the history of zero showed, ancients had a phobia about voids. Negative numbers were commensurately creepy.

The pox was not just on negative numbers. Positive square roots that did not neatly resolve were insufferable to mathematicians in medieval times.

Cubic equations were known to the ancient Babylonians, Greeks, Chinese, Indians, and Egyptians. Cubics arise in considering periodic occurrences, such as the orbits of planets.

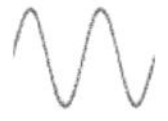

Cubic equations when graphed render a sine curve. Figuring the solutions to cubic equations (e.g., $x^3 + a_1x^2 + a_2x + a_3 = 0$) were long a source of consternation to mathematicians. Even depressed cubics, which lack the $x^2$ term, caused conniptions.

The struggle with cubics reached polymath Gerolamo Cardano in the mid-6th century BF. Cardano managed the feat. In doing so he encountered $\sqrt{-1}$. He grasped it. Though there were hiccups, others followed, and furthered the sense that $\sqrt{-1}$ was a workable concept.

The square roots of negative numbers were initially termed *sophisticated* or *subtle*. Then along came 5th-century BF philosopher René Descartes, whose frustrated failure to make geometric sense had him derisively call these numbers *imaginary* in 463 BF. Descartes was influential. The name stuck.

Cardano was first to combine imaginary and real numbers, creating *complex numbers*. He expressed them in the form $a + bi$, where $a$ & $b$ are real, and $i$ is the imaginary unit ($\sqrt{-1}$).

Geometrically, complex numbers form a plane. On one axis run real numbers, while the other axis is imaginary.

The man who tamed complex numbers was not a mathematician, but a surveyor: Caspar Wessel, a Norwegian. His brilliant paper on the subject was published in Danish in 301 BF. It made no impact in the mathematical world until rediscovered nearly a century later, in 205 BF. By that time, others had trod the same path.

Wessel's insight was to imagine complex number points within a 2-dimensional coordinate system, with real and imaginary number axes, as shown.

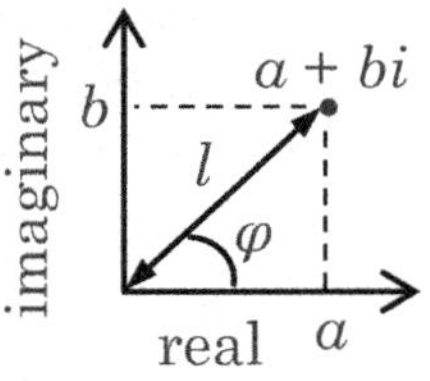


Next, consider a point ($a + bi$) as being at some length ($l$) from the origin (0,0), and at some angle ($\varphi$) from the real number axis. This puts a complex number point in polar form.

Using the polar coordinate system, each point can be considered as a distance ($l$) from a fixed point (typically (0,0)) at an angle ($\varphi$ from a fixed direction (e.g., the horizontal (real) axis)). Hence, the point $a + bi$ can be stated as (l, $\varphi$).

Using this scheme, mathematical operations on complex numbers are greatly simplified. Multiplying by $\sqrt{-1}$ is, geometrically, simply a 90° rotation counterclockwise.

Once made workable, complex numbers became essential for science and engineering. Complex numbers were necessary to model Nature: the realm of ultimate intricacy. Modern physics was only possible using complex numbers. Real numbers alone cannot describe cosmic or quantum mechanics.

Zero and $\sqrt{-1}$ illustrate that mathematics is discovery, not invention. This corresponds with the broader observation that perception is of a natural order.

Nature commonly exhibits self-organization and hidden order within apparent disorder. You can see this in ripples of water on a river. Chaotically tumultuous fluids spontaneously create stripes of coherent flow alternating with turbulent regions. Liquids self-organize into crystalline structures. Photons in laser light exhibit fractal patterns. Viewed as particles in a system (instead of linearly), prime numbers are an ordered structure.

The studies of cosmology, physics, chemistry, and biology all aver a natural order. Math highlights this. Comprehending Nature through mathematics cannot be invention.

Further, the patterns which the mind perceives convey an inner order. This natural order is not of Nature per se, but how the mind organizes perceptions to make them sensible.

Mathematics reigned over many disciplines of modern Earther life. Physics and economics were readily reduced to mathematical inquiries.

Hypotheses become theories by empirical buttressing which took mathematical form in models. "Mathematics is not just a tool by means of which phenomena can be calculated. It is the main source of concepts and principles by which new theories can be created," wrote 2nd-century-BF theoretical physicist Freeman Dyson. Conversely, existing theories were crushed by the weight of numerics.

Oddly, despite having proven otherwise, most Earther physicists made the mistake of thinking their formulas represented a 'real' physical world. The quantum physics models taken as reflective of Nature were simple sketches of deception. Naïve gullibility led physicists astray.

Economics was an empire of numbers, built on a foundation of exploitation. At the theoretical level, capitalism's copious failures owed to not numerically encapsulating factors critical to the well-being of societies. If the actual math of capitalism had been known, it would have been discarded centuries before the fall as untenable.

❀❀❀

Why Earthers went extinct is revealed by comprehending their mental mechanics. To begin, their predisposition toward materialism may be understood by looking at their developmental psychology.

In pure awareness, a human newborn was one with what it experienced. This sense of unity quickly evaporated as the mind came to life.

As with all altricial animals, what stuck in the mind of an infant was an inborn bond with mother. A neonate regarded its mother as unique. An infant sought out her mother's smell and face over those of other women. For those well cared for by their mothers, that bond remained strong throughout life.

By 2 months, babies began to think that they were confined within their own bodies amidst an external world.

At 8 months, infants began to grasp the concept of loss.

By 12 months, they started to form attachments to comfort-objects like blankets, which provided a temporary substitute for their mother. Around the same time, children started to speak their first words.

By 21 months, a word surfaced that became part of life's regular playlist: *mine*. "Children include ownership as an attribute of their object representations," noted 1st-century-BF psychologists Peter Blake & Paul Harris.

2-year-olds discovered frustration: that circumstances may thwart their wishes. Contemporaneously, the already well-developed propensity toward possessions sharpened.

Frustration underlined that possession cannot be taken for granted. 2-year-olds fought harder for toys they felt they owned. "Disputes over property are among the earliest, most frequent, and most intense conflicts in childhood," wrote 1st-century-BF psychologists Charles Kalish & Craig Anderson.

From 3 years on, ownership became a compelling personal and social force. A 3-year-old would protest if someone tried to take away someone else's toy.

"Materialism is a value-based response to insecurity in one's life," observed 1st-century-BF business academic Aric Rindfleisch. The degree of insecurity was a cultural trait. 60% of Western children were inseparable from their favorite object.

Strong object attachment in Western babies stemmed from their being separated from their mother while sleeping. Blankets and stuffed animals substituted for the intimacy they needed. Clinging to possessions among Japanese infants was rare, as infants slept with their mothers into middle childhood.

People valued their own possessions over similar items. Simply selecting an object endowed it with more worth that an identical one that went unpicked. People with a $1 lottery ticket they selected would only sell it back for $8, on average, compared to the $2 demanded for returning a ticket handed to them. This bias, felt from infancy, is the *endowment effect.* Other simians also felt the sentimental pull of ownership.

Materialism also served to state personal identity. "Our possessions are a major contributor to and reflection of our identities," wrote 1st-century-BF Canadian scholar Russell Belk, who echoed a common conception.

Objects as symbols of identity were the basis for brand-name products. Corporations lucratively relied upon such symbolic attachment.

The Japanese and Chinese especially esteemed branded luxury goods as identity statements. Flaunting such items was considered an acceptable way to stand out in a culture which espoused communalism as an ideal.

2nd-century-BF American philosopher John Searle termed cherished ownership a *status function.* Status functions establish an identity and provide a reason to act in a certain way. A status function is a mental attachment to an object by ascribing meaning, purpose, or function.

A subjectively perceived status function is a *personal fact.* When a status function is shared, it becomes a *social fact.* Money as a means for obtaining ownership is a status function that only works because it is a social fact.

Not all shared status functions were grounded as social facts. People intuited some statuses as natural, beyond any need for consensus.

From early childhood, people perceived morals as objective. Integral to this mindset was belief in a *moral universe*: a cosmic order imbued with immanent justice. Immanent justice is the idea that Nature enforces morality: that bad people will eventually experience misfortune and good people will receive good fortune.

Moralizing religions exploited people's naïve sense of a moral universe. The Hindu and Buddhist conceptions of *karma* – that people eventually got what they deserved – relied on this inborn belief in immanent justice.

Judaism also preached karma. "Tell the righteous that it shall be well with them, for they shall eat the fruits of their deeds. Woe to the wicked! It shall be ill with him, for what his hands have done shall be done to him." So said the evangelical prophet Isaiah in the Hebrew *Bible*.

Christianity further enfolded immanent justice in teaching that God judges individuals when they die: letting souls ascend to Heaven or be damned to Hell depending upon their acts and beliefs while alive. "For the wages of sin is death; but the gift of God is eternal life through Jesus Christ our Lord," wrote the apostle Paul in a letter which appears in the Romans chapter of the Christian *Bible*.

¤✧¤

Reverence for the possession of objects was the emotive basis for capitalism, which was founded upon private property as a fundamental right. "The idea that we make things a part of self by creating or altering them appears to be a universal human belief. The maker of an object, the user of land, and the cultivator of a plant are regarded as being entitled to the product of their labor," commented Russell Belk.

Belk was expressing what economists called the *labor theory of value*. "Commodities possessing utility derive their exchangeable value from 2 sources: from their scarcity, and from the quantity of labour required to obtain them," claimed 3rd-century-BF English economist David Ricardo.

Ricardo was echoed a half-century later by Prussian economist Karl Marx, who used the labor theory of value to argue for *socialism*: a political economy which served society, not individuals. "The values of commodities are directly the labour employed in their production," Marx wrote.

If labor was the source of economic value, workers should reap the full rewards for their efforts, not go to those who bet on investments, Marx figured. Marx used labor value theory to make his moral argument for equity.

To consider apportioning material objects or resources as fair or unfair is the moral refraction of materialism. Inequality as inequity was the moral ground from which opponents to capitalism fought. "Capital realizes value while labor creates value," Marx opined.

❀❀❀

There are 3 possible paradigms of reality. Knowing the correct one makes a profound difference in how life is lived. The Collective hewing to the wrong paradigms was instrumental in Earthers' self-destruction.

The human mind presented actuality as a duality. Their minds insisted that they were a mind-body amid an external world. The duality was of inner (mental) and outer (physical) worlds.

To natural philosophers since antiquity, dualism presented an unsolvable paradox: what is the interface between the mind and the body? How to reconcile physicality with mentality? No answer was ever found because there isn't one. The intractability of the 'mind-body problem' naturally led to a paradigmatic choice of a monism.

A *monism* posits that existence emanates from a single source: either the mind is made by matter (matterism) or matter is made by the mind (energyism). Matterism and energyism are diametric. Only one can be correct.

Nature is the perceived exhibition of existence. Its study is *natural philosophy*. As medieval times were drawing to a close, thoughtful men sought to shed the wooliness woven into natural philosophy since antiquity. They conceived *science* as an empirical quest for truth.

The Scientific Revolution began in Europe in the 6th century BF. The marked start was the 557 BF publication of *On the Revolutions of the Heavenly Spheres* by German astronomer Nicolaus Copernicus. The book was a geometric speculation on the orbits of the planets around the Sun. The Revolution's marked terminus was the 413 BF physics book *Mathematical Principles of Natural Philosophy* by English natural philosopher Isaac Newton.

The supposition behind both treatises was that Nature could be codified mathematically. Behind this assumption lurked the unexamined belief that there are laws of Nature. Science's unstated purpose was discovering these mechanistic laws, which were expressed as formulas where possible.

A key axiom of Earther science was that Nature is *objective*: existing regardless of perception. This falsity of this assumption should have been obvious, but apparently not to scientists.

In the early 4th century BF, British philosopher George Berkeley argued that perception is the only vehicle by which Nature may be apprehended. "Existence is perception. All those bodies which compose the frame of the world have not any subsistence without a mind," Berkeley wrote. Berkeley argued that existence is purely a subjective experience.

Berkeley's insistence that objectivity is nothing more than consensus was derided by contemporaries. Critics concurred that existence was objective. This objectivity occurred simply because they all agreed that it was so: a circular irony which made Berkeley's point.

Modern physics proved Berkeley right. Theories of both quantum and cosmic mechanics depend upon a subjective "reference frame" to explain phenomena. Still, the consensus remained that the physical world was mystically 'real' and thereby objective.

The apprehension of Nature is always perceptual, always a subjective construal. All one ever knows firsthand is what one directly experiences. All else is rumor. Though universally ignored, this simple tenet – *solipsism* – is undeniable.

As derision for Berkeley showed, the concept of objectivity is just concurrence on hearsay: that people agree to believe what others report. Gullibility goes down easiest, of course, when it corresponds with one's own conception.

All scientific arguments were disagreements over versions of objectivity. If instead one accepts solipsism, such arguments vanish, as all tales of Nature are treated as personal facts, with social facts coincidental.

The methodological glue that held science together was the *scientific method*. The scientific method was a slight twist on the hoary method of inquiry advocated by 27th-century

BF Athenian philosopher Socrates. Both methods aimed to eliminate false hypotheses. The conceptual implements to do so were inference and logic. Inference is making guesses based on observations. Logic is reasoning about relations and causes.

Look at the root of reason to see it for the ruse that it is. Reasoning is a train of thought which is found self-satisfying. "Though appearing to be intelligent, thought is unable to comprehend anything really," sagely wrote Indian guru Vasistha in the 42nd century BF.

The slight distinction in the scientific method from Socrates' technique was emphasizing evidence as a nullifier or promoter to elevate a hypothesis to a theory. A *hypothesis* is a guess. A *theory* is a guess seemingly amply supported by observations. *Evidence* is a construal turned into a social fact.

The social essentiality to science was fluidity. Science was mere dogma if disproven theories were not discarded. Such was the stagnation of modern Earther science, as supposition on the universe illustrates.

Late-century cosmology was ruined by a single assumption: that the first detectable light signaled the beginning of the universe, 13.8 BBF. Many observations belied this premise: most prominently, mature galaxies just after the cosmos supposedly started. Yet astrophysicists insisted upon this marker, as it was the only evidence of origination.

That the cosmos commenced with an indelible statement was silly: quite unscientific. Yet cosmologists clung to it. This refusal to toss an obviously bad assumption owed to institutionalized groupthink. Earther astrophysics in the last century was a joke of collective delusion. A dominant dynamic of Earther civilization in its last stage was institutionalization, which acted as inertia to progress in both science and society.

With regard to the social facts, cliques of power radiated what principles were acceptable. After their dismissal, dissenting views were shut out. Thus it was in teaching sciences at modern universities, from which the public learned what should be considered valid. "Truth is the cry of all, but the game of the few," observed George Berkeley. The ruling confederacy of Earther scientists tacitly agreed that their Collective construals directly reflected reality. The public naïvely went along.

Earther scientists prized evidence above all else. Consensus about artifacts was construed as constituting an objective reality. As such, science theories were invariably cast in terms of matterism.

Earther scientific gospel was that matter made the mind. The irony was that discoveries blatantly exposed the falsity of matterism. Scientists simply ignored the abundant disproof of matterism, refusing to consider the obvious implications of incontrovertible evidence on every front. The most generous assessment of matterist scientists was that they were sophisticated simpletons who conflated correlation with causality. They confused the appearance of things with their essential nature: the root of naïve realism.

Grounded in matterism, Earther 'science' did not earn its claim. It was instead a shifty religion. Its popular credence owed to studied ignorance. The theoretical dissonance in evolutionary biology is illustrative.

In the mid-4th century BF, English naturalist David Hume remarked on evolution: "A purpose, an intention, a design, strikes everywhere even the careless, the most stupid thinker." Alas, that damning characterization was beyond most scientists.

In the early 3rd century BF, French evolutionary biologist Jean-Baptiste Lamarck pinned evolution as adaptive, based an organism's experiences. "Behavior may drive evolution. Everything that Nature has caused individuals to acquire or lose by the influence of circumstances, it preserves by heredity and passes on to the new individuals descended from it," Lamarck wrote. Spot on.

Lamarck was ignored in favor of English naturalist Charles Darwin, who clumsily fumbled about a half-century later with an assortment of guesses to explain evolution.

Darwin was a keen observer but a pathetic theoretician. His 1st take on evolution was that it is saltational, involving creative leaps in forging new species. Darwin later rejected saltation for a more stolid story.

Inspired by Charles Lyell's hypothesis of a gradually changing geophysical Earth, and Thomas Malthus' surmise of food supply limiting population growth, Darwin derived a notorious guess about biological evolution: gradual change in species based upon existential competition. Darwin's "*natural selection*" was concocted from the notion of there being a struggle for survival at the species level. He never explained how that happened.

Lacking any sense of biomechanics, Darwin speculated that species somehow sorted out hierarchically, based on competitive edge. Darwin espoused that "natural selection almost inevitably causes much extinction of the less improved forms of life." Channeling Lyell, Darwin imagined that natural selection was a slow progression.

In his later years, Darwin dabbled with pangenesis, which was a complex mishmash of heredity enveloping sexual reproduction, life experience, and homeostatic phenomena, such as cellular regeneration. Darwin's musings on pangenesis harkened back to Lamarck. Darwin never did come up with a decent explanation of evolution.

What Darwin did have were influential promoters. The nebulous slogan "natural selection" became the buzzword for how evolution happened.

The discovery of DNA in the mid-2nd century BF opened an opportunity to explain evolutionary mechanics. Rather than figure in adaptation, geneticists promoted the silly idea that evolution transpired via random mutations.

In the last century, savvy evolutionary biologists came back to the idea that evolution was adaptive. Yet Earthers never did express a cogent theory on the mechanics of adaptation. They pointed to DNA and its chemical companions as causal, ignoring the simple scientific principle that artifacts can't do anything. To the very end, what Earthers believed as science was instead myths of magic.

¤✧¤

How adaptation happens is revealed in the next section. It is only half of the story of how existence is fabricated. The other half is filled in by physics, as described next. As to *why* existence is (as it is) can be gleaned from psychology. The 'why' story completes the context of the machinations behind Nature. That ground is covered before this chapter concludes.

A universal, unified field of cöherence localizes to create the perceived fabric of existence. Some local fields animate life forms, providing the minds which apprehend information from other localized cöherence fields – whence the show called Nature.

Local cöherence fields evolve life as adaptive proposals aimed at succoring survival. Adaptations may not work out because habitats change. The challenge of living is ongoing.

Advances in physics in the early 2nd century BF provided the potential to create a holistic theory of reality. The founder of quantum physics, Max Planck, set the stage. He was soon joined by maverick Albert Einstein. After that, theoretical physicists faltered and failed.

In 200 BF, German physicist Max Planck theorized that Nature was quantized: its appearance invariably broken into bits. Experiments verified this.

Beyond the fracturing of matter, Planck comprehended the essentiality of consciousness, writing: "I regard consciousness as fundamental. I regard matter as derivative from consciousness. Everything that we regard as existing postulates consciousness."

Planck on cöherence: "All matter originates and exists only by virtue of a force. We must assume behind this force the existence of a conscious and intelligent Mind. This Mind is the matrix of all matter."

Albert Einstein was inspired by Planck's quantization discovery and emphasis on consciousness. Extending from it, Einstein concocted his theories of relativity.

Einstein's relativity pivoted on Berkeley's insistence that perception was key to existence manifesting. The experience of Nature depends upon a witnessing consciousness. It was an obvious observation which most physicists ignored, both before and after Einstein. "Physical concepts are free creations of the human mind, and are not, however it may seem, uniquely determined by the external world," wrote Einstein.

Isaac Newton had been adamant that space and time were "absolute." But space and time are mental fabrications, and therefore relative to individual perception. The great revelation of modern physics was that the supposed absolutes of classical physics were instead precepts of perception.

The criticality of consciousness was apparent in quantum physics. The appearance of quanta owed entirely to their perception. "There is no quantum world. There is only an abstract description," noted 2nd-century-BF atomic physicist Niels Bohr.

In 195 BF, Einstein theorized that matter was made of energy (the famous equation: $E = mc^2$). 4 decades later, that point was explosively proven by atomic bombs.

Earther physicists did not appreciate the implications of matter/energy equivalence. If they had, energyism would have been adopted as the dominant reality paradigm. It never was.

That matter is energy was indisputable to modern physicists. But the key insight of the tautology did not sink in.

The idea of energy was exactly that: an idea. "Energy is an abstract concept introduced by physicists to better understand how Nature operates. Since it is an abstract idea, we cannot form a concrete picture of it in our minds," wrote 1st-century-BF physicist Carlos Calle.

"Nothing happens until something moves," remarked Einstein. The animation of matter – energy – is the wellspring of existence. Just as perpetual motion machines are impossible, matter cannot move itself.

Beyond its animation, matter itself is made of energy. Energy is only an idea. That necessarily means that matter is a mirage made by the mind. As matter is of energy, matter must be an artifact created by the mind from data provided by cöherence.

The mind's contrived artifacts are an artifice. To think of substance as real is to be deluded. The mind convincing its occupant consciousness that matter is authentic is the primordial deception.

Conventional Earther conception was that perception captures a material world in the mind: a turning of matter into symbols. The opposite is true. Perception converts symbols into the appearance of matter.

Nature is an intricate realm of symbolic representations. The world emanates from a unified data plane instantaneously collated via localized fields of cöherence. Energy is data by which Nature emerges moment-by-moment in the conscious minds of its perceivers.

¤✧¤

The vision systems of all of Earth's organisms have been attuned to the narrow solar bandwidth richest in information content: *light*. That adaptation was no random accident. Other spectral bands, such as microwaves or X-rays, are comparatively paltry in their proffered data.

Earther physicists were long fascinated with light: the mystical illumination that made the creations of cöherence visible.

In 2140 BF, Hero of Alexandria noted that light takes the shortest path to its destination. Light refracts to do this.

Light travels slower in water than air. To ease its way, a beam of light refracts when it hits water, thereby optimizing its travels. Similarly, light from the Sun repeatedly refracts when traversing Earth's atmosphere, which is denser than the relative void of outer space.

In the millennia that followed Hero, refraction was recognized as common to all energetic rays. Radiative transmission optimality was generalized and incorporated into all physics models.

Oddly, the direct implication of this went unremarked. Optimal propagation (as exhibited by refraction) is only possible when all information about existence is instantaneously known. This implies a unified, coherent intelligence from the quantum level on up, and indicates *teleology*: a guided presentation of Nature.

¤✧¤

In 136 BF, English physicist Peter Higgs theorized that the character of quanta derived from a universal field which localized. The *Higgs mechanism* became universally accepted among physicists, as it was consistent with all observations, and provided the conceptual means for consistency in the supposed laws of Nature.

In a simultaneous process of localization and collation, a universal field of cöherence mystically provides comprehensible perception. “The most incomprehensible thing about the world is that it is comprehensible,” marveled Einstein.

Like cöherence, a universal field of Cönsciousness localizes to populate every life form with consciousnesses. Vedic texts written in the 35th century BF taught this principle. In that tradition, 2nd-century-BF guru Nisargadatta Maharaj summarized: “It is not really the individual that has consciousness. It is instead that Cönsciousness assumes innumerable forms.”

The ö in Cönsciousness and cöherence signifies localized diversity (the 2 dots) from unity (the circle).

Though acknowledging the wiliness of viruses and cells in science journals, and despite pointers from Planck and Einstein, Cönsciousness localizing was not accepted by Earther scientists. Instead, intelligence was attributed to animal brains and nerve cells: the matterist delusion. Flummoxed Earther biologists could not explain how brainless life, such as plants, could be aware of their environment and act intelligently.

That matter is made of energy and energy is data are the fundamental truths that Earther scientists discovered but did not comprehend. The very idea of matterism was nonsensical. Such was the state of Earther ‘science’ as men extinguished their species.

¤✧¤

For every living entity, from viruses to humans, the mind fabricates the show called Nature. This production is made to consciousness.

Consciousness is the faculty for awareness. Consciousness witnesses what the mind presents. “Consciousness is not something we see. It’s something through which we see,” explained 1st-century-BF biologist George Mashour.

Human consciousness had states and levels. States were periods of awareness, such as awake or asleep. Levels were planes of clarity (in awareness). Consciousness levels afforded distinctive perceptivity.

Cönsciousness is the fundamental force (unified field) of awareness which provides the ability to witness the creations of cöherence.

Cöherence is the universal field which manufactures minds and the contents they consume. Cöherence is the creative force behind existence.

Cönsciousness and cöherence are spoken of as if they are separate. That distinction is just a device to facilitate understanding. Cönsciousness and cöherence are unified: the monism from which duality proliferates.

The evidentiary basis and theoretical underpinnings to understand that matter was merely a product of perception was incontrovertible. Yet, despite millennia of instruction from revered sages, the principles of energyism eluded ordinary Earthers. Instead, the Collective stolidly accepted dualism *and* matterism without distinguishing between the two.

The term *Collective* refers to the 99.9+% of humanity who stewed in iğnorance: the lowest level of consciousness. Those in iğnorance suffered mental illness: floating fears, worry, anxiety, depression. This inner dysfunctionality occurred because people gullibly believed their mind's imaginative productions.

In iğnorance, awareness is dimmed by an incessant inner din. This noise is generated by a seemingly independent mental agent: *nattermind*. Iğnorance was slavery to nattermind. The breve symbol ˘ (over ğ) signifies nattermind lording over the mind of those in iğnorance.

That the Collective failed to appreciate that *level* of awareness was critical to quality of life spoke volumes about their existence, and their inexorable demise.

"The ultimate value of life depends upon awareness," emphasized 26th-century BF Athenian natural philosopher Aristotle. Deluded by their minds, those in the Collective ignored this crucial truth. Instead, when feeling stressed, they found solace in stupor: self-medicating with chemical agents that furthered lowered awareness.

The upshot of iğnorance was ignorance: accepting falsities as truths, and from fantasies building belief systems which served as obstacles to healthy living. A 31st-century-BF Vedic text, *Katha Upanishad*, summarized the Collective effect: "Fools dwelling in iğnorance, yet imagining themselves wise and learned, go round and round in crooked ways, like the blind led by the blind."

❀❀❀

The will to live is a prerequisite for complex life to evolve. Earth's eukaryotes, upon which all multicellular life was based, owed to *symbiosis*: the joining of cells by way of trust.

The root of that trust was will to live. If cells readily surrendered their lives when stressed, they would not be reliable partners. Resilience is integral to reliability.

Biology is built on survival. Will to live is the universal thrust of *vitalism*: the vital life force. Striving for survival is the impetus behind adaptation.

Psychologically abetting the drive to survive is optimism. This was notably true in humans: a creature which believed it could craft its own fate. This can-do confidence was collectively reinforced by the technological leaps men made in their last few centuries.

This was ironic. Technology precipitated self-extinction. Yet Earthers believed themselves a superior species, and practically invincible, because of their technologies.

As a natural extension of will to live, the urge to biologically reproduce compels all organisms. For women, a want of intimate nurturing was sufficient. Men envisioned fatherhood as a form of immortality.

What compels the will to live is simple: enjoyment. Every life desires another moment of the sheer pleasure of feeling alive. When that will is lost, death inexorably follows.

⌑✧⌑

Nature is a show created by cöherence, serving as an entertainment venue for Cönsciousness in its innumerable incarnations. That platform is designed to reward living life to its fullest. This design parameter explains why living is a challenging puzzle.

The richness of Nature naturally inspires wonder and awe. This easy amusement is exceeded by gratification.

The deepest pleasure is the satisfaction of accomplishment: of overcoming obstacles to produce one's own works.

Earther culture was at its epitome celebrating creativity, whether functional or aesthetic.

The lynchpin to accomplishment is struggle. The outcome of effortless action may be pleasing, but it is not a triumph.

Gratification emanates from focus: awareness and comprehending a conceptual system from which productive effort emerges. Physical activity may be instrumental, but it is incidental to the glow of fulfillment. Fun is in the mind.

The mind also generates dissatisfaction. Nattermind soured disillusionment into mental illness.

Therein lie an incentive. Grating discomfort was the spur to transcend the mind's predations.

The reason for iğnorance was to maximize entertainment potential. To this end, the human mind had a built-in disability component (nattermind) which could be overcome, but only through resolute effort. Elevating consciousness was the most exquisite accomplishment possible.

Techniques for purifying awareness had been taught since prehistory. Yet few exercised the will to realization: whence the lazy-minded Collective and the dynamics which ushered humanity's fate.

Through its incessant and insistent commentary, nattermind was the agent which created iğnorance. The invariable reckoning to nattermind's beckoning was mental disability.

Most in the Collective ignorantly thought mental illness normal: simply something that was part of living. Collective envisionings of enlightenment were cartoons of the pure awareness that transpires in unity consciousness. The paradigm shift from iğnorance to self-realization was beyond what those in the Collective could even imagine.

Moderns maintained institutions succoring iğnorance. The mainstream press sensationally touted mental illness as a disability with benefits, such as a source of creativity. This celebration of inner sickness goes a long way in explaining Earther's terminal predicament.

The only cure to nattermind's pollution was inner silence. Subduing nattermind afforded full awareness to the present moment, which is all that ever exists. All else is the mind presenting a mix of memory and imagination.

Abiding inner quietude characterizes the *enlightenment* level of awareness. Continuing elevation of consciousness reaches an apex with *unity consciousness*, also known as *realization*.

In unity consciousness, a person viscerally experienced that one's own consciousness is of the universal field. In realization, consciousness witnesses the mind's fabrications.

Insights into reality abound in realization. That is why gurus embodied wisdom. A *guru* was a spiritual teacher in realization.

The term *realization* came from archaic Sanskrit. That word could be found in the earliest Vedic texts of Hinduism, dating to the 37th century BF. Arising in India, Hinduism was the oldest historical religious creed. Most gurus in Earther history were steeped in the Hindu tradition.

The ascension from iğnorance to realization was Earthers most difficult challenge, and thereby afforded the greatest gratification. "The supreme calling of every human being is to aspire to self-realization," advised Indian guru Anandamayi Maa in the 2nd century BF. Though the power of this beckoning was taught by gurus from antiquity, moderns in the Collective were oblivious to it. They depreciated their own lives in iğnorance.

From cöherence-fed symbolic data the mind presents a tangible world of objects, and then transforms it back into abstraction. This is deceptive double-take processing.

Nature is an endless river of constantly changing novelty. Every moment and all perceived are unique. Yet uniqueness is an obstacle to comprehension.

Abstraction is necessary and necessarily false. Categories are both essential and nonexistent.

As everything is unique, perception is an economizing process. The inner vignettes of memory transpire through autonomic categorization. The mind is a sketch artist.

Mentation is patterning: symbolic representation leading to conceptualization. The process of learning is an expansion of concepts upon which experiences may be slotted and new skills built. Intertwined with this expansion is simplification.

Recognition is facilitated by remembering only rough portrayals. People with photographic memories have trouble recognizing faces because features are distinct depending upon lighting and viewed angle.

Recall is a mixture of memory and imagination. Imagination fills in details from the sketches of remembered events.

More creatively, imagination manufactures perceptions of events which did not happen or scenarios of what may be. Imagination is an extension of conceptualization: a paracosmic projection of symbols into fabricated patterns.

Imagination is essential to problem-solving: envisioning what is not to achieve what may be. Viruses and microbes solve problems. The distinction in hominids was the degree to which the imagination was set loose, and the esteem with which it was held. Cultures, religions, and ideologies were imagined castles built on the sands of abstractions. Figments were forged into convictions among the Collective via vivid narratives.

Animals hew to actuality because their lives depend upon it. Hominins evolved to be an exception. The mental power that Earthers drew upon was the summoning of fantasy. It was a facility fraught with peril.

Many animals have used and crafted tools, built homes, communicated via languages, played, and practiced culture. Humans took all these things to flamboyant extremes, creating mental reference frames abstracted away from Nature.

People increasingly became taken with their paracosms, which ironically evolved into the stabilizers of societies. These fruitions came late in Earther evolutionary descent.

Hominids expressed representational art for more than a million years. Such imagery became more elaborated, and appeared in new materials, with technological advance.

Hominin craftwork expressing abstractions only began to proliferate 20,000 years before the fall, during the period known as the Mesolithic. This was the final era of foraging, before agriculture began its dominance. Religions appeared much later.

⌑✧⌑

Belief was a potent force in Earther affairs. Practicality was a pittance compared to belief, even in modern cultures. Understanding the nature of belief unravels the spell it casts.

A *belief* is confidence in an abstraction as truth. Belief is an ossification of imagination.

At the core of every belief is a *value construct*: a concept of cherished meaning. The invariable reason for belief is for mental comfort. Beliefs succor a sense of surety about what otherwise might be disturbing.

"A belief is not merely an idea that the mind possesses; it is an idea that possesses the mind," wrote English scholar Robert Bolton in the 4th century BF. "Once you have a belief, it influences how you perceive all other relevant information," furthered American political scientist Robert Jervis 3 centuries later.

Paracosms as cherished belief systems were an Earther trademark. The Collective were in thrall to their beliefs.

❀❀❀

The earliest societies sprouted in India and China. Their social cohesion came through the teachings of gurus.

Further, gurus originated all major Earther religions. Their lessons were invariably distorted by lesser men, who institutionalized religions into political control mechanisms.

From their initiation in the 37th century BF, the Vedas were the earliest catalog of writings by the gurus in India. Though inspirational to spiritual seekers, these texts were esoteric: often relying upon parable and seldom straightforward.

A central conceptualization of Hinduism was *brahman*: an infinite, eternal, transcendent force signifying absolute reality. Brahman was a reference to the unified field (monism) of Cönsciousness/cöherence: Ĉocö. The concept of Ĉocö was disfigured into a supreme being when Hinduism morphed into a moralizing monotheism.

Lao Tzu was a legendary Chinese guru who lived in the 27th century BF. The name itself is an honorific title meaning *venerable master*. Next to nothing is known about Lao Tzu as a person. Some think he lived before Confucius (2651–

2579 BF). Some think he lived later. Some think he was Confucius.

Confucius was a social philosopher whose teachings wove the philosophical fabric of China's civilization throughout its history. Confucianism emphasized rectitude, fairness, and beneficence in all relations.

Confucius' socialist suggestions coincided with those of Lao Tzu. But Lao Tzu's writings dove deeper. The spiritual legacy of Lao Tzu was Taoism, which infused Chinese tradition with an emphasis on naturalness, simplicity, and harmony.

Lao Tzu wrote of the Tao as a unified yet diversified field: Ĉocö. "The great Tao extends everywhere: so formless, so intangible. When the Tao is expressed, it seems without substance. The Tao is empty and yet useful. It resembles the source of all things.

"The Tao acts as natural law. All things depend upon the Tao for growth, and it does not deny them."

Lao Tzu wrote of realization and how to reach it. "Those who master others have force. Those who master themselves have strength. Cultivate the inner self and its power becomes real. Clarity and stillness bring order to the world.

"One may travel far and know little. Those realized know without going about, recognize without looking, achieve without acting.

"One in realization has no fixed mind. Those skillful in the Tao are not obvious to the Collective. They appear to be simple-minded."

Lao Tzu appreciated that those in the Collective who do not seek enlightenment consider the endeavor farcical. "My words are easy to know, easy to follow. Yet the Collective are unable to know them, unable to follow them.

"When superior people hear of the Tao, they diligently try to practice it. When the average person hears of the Tao, they appear both aware and unaware of it. When inferior folk hear of the Tao, they roar with laughter."

Whereas a guru was a realized teacher – such as Lao Tzu – a prophet was someone with a vision of the future. Few Earthers possessed both.

Historically, those realized or with the gift of prophecy were either revered or thought insane, depending on their circumstance. "To the iğnorant, all that they cannot understand is madness," observed Maharaj.

The year was 2166 BF: 4 years before the Romans rained ruin upon Jewish Jerusalem. Jesus ben Ananias wandered the city, prophesying that it would be destroyed.

Jewish leaders, annoyed with Ananias disturbing the peace, turned him over to the Romans for prosecution. Ananias was tortured, but then released as a madman because he had shown no concern for his fate while being tortured.

Ananias persisted with his prophecy until killed by a catapulted stone during the Roman Siege of Jerusalem in 2170 BF. In the moments before he was struck and killed, Ananias uttered, "Woe once more to the city and to the people and to the temple, and woe to me also."

Ananias' bitter prophecy proved true. The siege ended with the city sacked, the destruction of Jerusalem's famed Second Temple, 1.1 million Jews killed, and 97,000 enslaved.

Less than a century later, another Jesus paid the price for annoying the authorities. Jesus of Nazareth lived in the 21st century BF. It was a time when the Jews looked for a prophet to lead them from Roman oppression.

But Jesus was no prophet, nor much of a guru. In his short life, Jesus managed to corral only a dozen or so disciples, as well as a wife. At the age of 33, irascible Jesus got himself crucified by the authorities for rabble-rousing.

Jesus left behind no writings. His small retinue of followers admitted they never did comprehend what Jesus tried to teach them. Recondite would be an understatement. But crafty manipulators transformed Jesus into the son of God.

For decades after Jesus' death, a small, devoted group of Jewish mystics spun tales of Jesus. Later fables became increasingly far-fetched. The most notable of these fantasies was that Jesus resurrected from the dead.

Jesus' postmortem followers splintered into sects in the 20th century BF. By the end of that century, a single sect had become dominant, by dint of spinning appealing dogma.

This sect was far from Jesus' teachings of spiritual self-realization. In step with Jewish tradition, the sect emphasized moralizing monotheism, along with emphasis on faith and orthodoxy as the path to fabled eternal life. To believe and obey was enough.

Touting a superhero savior, the ruse worked. The Catholic Church's political ascent was assured in the 19th century BF, when Roman emperor Constantine 1 adopted Catholicism as the state religion. The Catholic Church was a potent political force throughout Europe during medieval times.

Power corrupts, as it did those leading the Catholic Church. A marked schism in the 6th-century BF, known as the Protestant Reformation, weakened the Catholic Church. But Jesus' legacy lived on. Nearly 1/3rd of the world's peoples counted themselves as Christians when Earther civilization collapsed.

The most influential guru-prophet was Muhammad ibn Abdullah, who lived in the 16th century BF. Muhammad unified fractious Arab tribes, and then served as both moral and political leader until his death. Muhammad left behind copious texts on a wide variety of subjects.

The inspiration of Muhammad resulted in Islam becoming the most advanced civilization of its time. This progress came to a sudden halt at the beginning of the 10th century BF, when conservative clerics gained power.

Islam regressed in the centuries that followed. Acting as a degenerate social disease, conservatism gripped Islam until the fall. The Muslim world's only salvation, and then only economically, was the discovery of vast reservoirs of oil under their lands in the early 2nd century BF, as industrialization was burgeoning. This viscous black gold, along with coal, were the fossil fuels which powered Earthers' surge to self-extinction.

In the 1st century BF, the number of Muslims was 2nd only to Christians. 25% of the world population was Islamic.

The orthodoxy of Christianity and Islam were moralizing dogmas that did nothing to propel their adherents toward mental health. Instead, there was a correlation between religious fervency and disturbing mental illness, including extreme violence and sex abuse.

Whatever consolation it may occasionally offer, belief is itself the crux of mental illness. Healthy people had no need for such inner crutches.

"God is only an idea in your mind. Give up all this trash, whatever you are studying in the name of religion, in the name of spirituality. Understand only one thing: that consciousness at present is your nature – you are that only," Maharaj advised. Few knew such wisdom, and even fewer heeded that counsel.

❀❀❀

In Earthers' last centuries, after science had become a religion, the Collective worldview strangely straddled the irreconcilable philosophical divide between naturalism and supernationalism. Oblivious to a fault, few in the Collective even recognized that such a chasm existed.

"Religion – or the belief in an invisible, supernatural order – exists in all human societies," wrote 2nd-century-BF political analyst Francis Fukuyama. Such widespread belief arose from fear.

Between the ages of 7 and 10 years, children fathomed that death is permanent, irreversible. This comprehension often led to lingering anxiety. Despite a growing ability to rationalize, fear of death commonly persisted into adulthood. Though ostensibly focused on living a righteous life, religion was, ultimately, a mortality ritual.

Every Earther religion promised consequence for bad acts: if not in this life, assuredly afterwards. Fear of retribution in the afterlife was the backstop of every religion. "If death were a release from everything, it would be a boon for the wicked," figured 26th-century BF Athenian philosopher Plato.

*Naturalism* is the belief that actuality and reality are synonymous: that observable Nature is all that there is. Naturalism dated to ancient Indian and Chinese philosophies which expounded *atomism*: that Nature consists of interacting atoms. Modern science also took this matterist stance.

*Supernaturalism* posits a discontinuity between actuality and reality. That gap created the philosophical space for religion to emerge. Supernaturalism splintered into several schools.

That Nature is a mystical, entangled unicity was a prehistoric apperception. From antiquity, energyism was taught by all true gurus.

In the modern era, energyism remained a tradition in some preindustrial groups. But tenets of energyism were seldom espoused in industrial nations. Only the obscure, last-century guru-scholar Ishi Nobu fully elucidated energyism.

Animism was an early supernatural conception: spirits were inherent in all matter, inanimate and animate. An optimistic variant of animism was that powerful spirits might be turned into temporary allies. This active animism arose from the hope that the capricious forces of Nature might be domesticated to some degree.

Animism independently evolved into polytheism among many cultures 7.5 TBF. Polytheism was belief in a plethora of gods. Polytheism independently emerged in the earliest political states, including Mesopotamia, Egypt, India, and China. "The most successful ancient empires all had strikingly non-moral high gods," noted 1st-century-BF anthropologist Nicolas Baumard.

An offshoot of polytheism was pantheism: the belief that reality was identical to divinity. Pantheism was a steppingstone to monotheism, and to the birth of moralizing religions.

Monotheism arose 5 TBF: first in China, then in India, and then among the Jews of the Levant, 28 centuries before the fall. Politics was behind this development.

Watchful gods arose in many mythologies before monotheism became the religious norm. Among other gods with supernatural sight, the ancient Egyptian sky god Horus was often depicted as a sharp-eyed falcon.

Monotheism took omniscience to a new level, as it was coupled with a single, all-powerful God: one who was supposedly a stickler for morals. This theistic package was devised for social control. "Mindful agents serve as a powerful source of social influence and control, increasing adherence to socially accepted norms of conduct, whether those others are actually present or merely presumed to be present," wrote social psychologists Nicholas Epley & Adam Waytz in the 1st century BF.

God did not arise until deemed useful by political authorities. Legal codes first emerged in the 5th millennium, contemporaneous with monotheism.

It was not until the Axial Age, between the 30th and 25th centuries BF, that moralizing ideologies gelled across the world. The idea of a judgmental God was sold by the dominant tribe to bind together the sizable societies which had emerged from the conquests that created them. "People say we need religion when what they really mean is we need police," observed 1st-century-BF scholar H.L. Mencken.

The evolution of moralizing religions reflected changes in societies and caused changes in those societies. Materialism and moralizing monotheism were the complementary sides of the same conceptual coin.

Prosperity provoked conservatism among dominant groups striving to hold their social positions. Moralizing religion was part of that package. "Affluence changed people's psychology, and, in turn, it changed their religion," remarked Nicolas Baumard.

Church and state intertwined in early civilizations, with pyramidal societies of hierarchical social castes. The vast majority lived as enslaved laborers: the base foundation of the social pyramid.

The idea of a "divine right of kings" spanned the ancient world. Christianity kept that idea alive in Europe until eroded by overweening materialism, brought on by burgeoning trade. Modernizing capitalism commoditized divinity, minting it into money. Wealth was worshipped.

In a tangible sense, plutocracy was Earthers' last religion. The United States of America was, constitutionally, a secular nation. Yet the state stamped on all its currency: "In God we trust."

¤✧¤

Organisms modulate their behavior according to their habitat. For instance, young starlings reared in a highly competitive environment invest less in their physical maintenance as adults. This lessens lifespan. Such starlings also develop a *fast psychology*: preferring immediate rewards over riskier, but potentially more profitable, foraging.

Humans were no different than other animals facing long odds of a long life. Those repressed also adopted a fast strategy: grow up quickly and have offspring early and close together, thus maximizing the probability of leaving some viable progeny. "Animals in a poor biological state face reduced life expectancy, and as a consequence make decisions that prioritize immediate survival and reproduction over long-term benefits," summarized 1st-century-BF ethologist Melissa Bateson.

Fast psychology had deep historical roots. Lao Tzu lived in the shadow of a powerful political state, ripe with moralizing and inequity. He warned of insidious societal dynamics. "When the great Tao is forgotten, philanthropy and morality appear. Stratagems are produced and great hypocrisies emerge. When the family has no harmony, piety and devotion appear. The nation is confused by chaos, and loyal patriots emerge."

For the elite who ruled a society and its religion, fast psychology was incipient subversion underfoot: brimming with potential for revolution. So – to foster taking it slow – religions came to condemn fast behaviors. Sex was religiously circumscribed to couples who had ritually pledged matrimonial fidelity. The promoted sole purpose of sex was procreation. Genitalia were considered obscene, as were women's nipples: though ample bosoms were universally esteemed.

Life-speed strategy may also explain the gradual decline of moralizing religion in wealthier, decadent societies as the fall neared. Such was the case in western Europe and North America. As more people (the middle class) become materially comfortable, moralizing religions become less relevant. Their adherents thinned.

That people also lost confidence in institutions abetted the decline of religious practice. Christian churches were repeatedly rocked by scandals. There was extensive pedophilia among Catholic priests, as well as its cover-up by clerical authorities. Making sex naughty turned it into a perversion. Those psychologically repressed by their beliefs were eager perverts.

People lost faith in institutions, but not in belief. Even in the last century, 90% of the Collective believed in a God.

Those in modern societies who still held to moralizing religions in the last century were largely the underclass: people who clung to traditions and feared change as a threat to holding onto what little they had. Clutching to religion also served as a supportive social badge, signifying an in-group by belief. Tribalism by religion appealed to those who felt beset.

❀❀❀

There was an odd dichotomy between what was known about Nature as contrasted to the false dogmas perpetuated as knowledge among the Collective. Crucial insights about reality had been the core of all religions. The lessons of gurus were corrupted for political advantage. In the modern era, their scientific rediscovery was ignored in favor of illogical explanations.

Concepts don't exist. They are capricious mental bundles: inner sketches of convenience. Believing an idea true ignores the uniqueness of everything experienced. As Maharaj noted, "All knowledge is conceptual, therefore untrue."

Faith in the mind's patterning is not a slippery slope into self-delusion. Instead, belief is itself the crippling fall that defines iğnorance. Acting through ideas versus believing in them is the difference between using a tool and being in thrall to it as a mystical signifier.

Such was the irony behind science. The practical benefits from scientific findings were multifarious and often profound. Yet, in embracing matterism, science was, at root, a religion. This was most telling in the science that mattered most: political economy, the science of societal well-being.

The significance of Collective iğnorance cannot be overstated. Fables which passed as truths diminished humanity and paved the road of self-destruction.

# The Evolution of Civilization

The global diaspora of hominids began early in their descent. They roamed throughout Africa and Asia for millions of years, speciating as they went. These creatures then traversed Europe and Oceania, making their way to the Americas as soon as passage became possible. By the advent of history, humanity's habitat was all Earth that a large land creature could claim as habitable, including many inhospitable places.

Hominids were a generalist omnivore and apex predator. No other large animal was better adapted to survive in a diversity of biomes.

Beyond any urge to be on the move, erratic climates prodded searching for richer lands. Beginning 31 millennia before the fall, a disparaging drier climate drove waves of hominids out of Africa.

❀❀❀

Hominids were not socially distinct from their ape predecessors. Family was the core of social organization. Mothers were the nexus.

Many creatures on Earth lived via fluid sociality. Among them were fish, cetaceans, ungulates, elephants, most mammalian carnivores, and primates.

Among hominins, a community of up to a few hundred individuals had dynamic subgroups that included families, friends, and coalitions. Subgroups merged for safety, such as for sleeping, or during conflicts with conspecifics. They then split up for foraging, or when disagreements arose within the larger group.

Families banded together through close social ties. Foraging clans were nomadic extended families. Reciprocity was the rule.

Mobile foragers were egalitarian. Such clans sometimes had elders and initiates, but no chiefs. No one commanded authority over adult males. Foragers leashed ambitious leaders.

Women typically had considerable social and sexual choice. Relations between the sexes varied, but women were an indispensable asset. Women gathered and captured food. They made much of the equipment that afforded comfortable living.

Compared to women, whatever archaic men did paled in practicality. Women were the ones who developed sewing, weaving, basketry, pottery, woodworking, and building. Owing to their social skills, women were the primary prehistoric traders.

The skills of a band of foragers, herders, or hunters were complete in their self-containment. They otherwise could not survive.

Knowledge was key. Its accumulation was the legacy of women. Mothers reared and educated offspring, molding them into productive members of the clan.

Women crafted a suitable place into a home. It was she who added man to her domesticated animals. Women weaved the comity essential to civilization.

Hunter-gatherers were committed to sharing. Many communities actively sought equality. Open-faced fairness was the only course by which group bonding could be sustained.

Foraging bands had nothing resembling later (modern) economic exchange, nor even the thought of individualism. The universal practice of eating together derived from the hoary practice of sharing food.

Many of the moral rules for clan society were not directed at transgressions against private property, but against *not* sharing. Facing the prospect of perpetual scarcity, failure to share can affect a group's survival prospects. Hence, early peoples had a heritage of egalitarianism. With rare exception, this practical equity was forsaken after food surpluses became possible.

As agriculture became more arduous, men took control of it. The leadership which women had naturally taken was wrested from them. "The application to agriculture of those very animals that woman had first domesticated led to her replacement by the male in the control of the fields; the advance from the hoe to the plough put a premium upon physical strength, and enabled the man to assert his supremacy," wrote Will Durant. The patriarchy which would forever diminish humanity sprouted when crops became an enterprise – when humanity became mankind.

Sexual activity as a moral issue only began with property regimes. "Marriage began as a form of the law of property, as a part of the institution of slavery," noted Durant.

Losing virginity was only a social stigma for girls. Condemning sexual activity as immoral applied only to women. In the crime of marital infidelity, the blame always laid with the temptress.

⌑ ✧ ⌑

Hominids established settlements long before they descended into humans. Like other motile animals, hominids first settled in places of natural abundance.

The sites of the earliest large societies were forested areas near fertile rivers. Those forests were invariably diminished by human habitation.

Cave paintings and fossil remains show that hunters had a strong preference for large-hoofed mammals. This preference remained unchanged for hundreds of thousands of years. 30,000 years before the fall, these meat supplies dramatically dwindled as overhunting took its toll. Men had a chronic habit of depleting their habitats in every aspect.

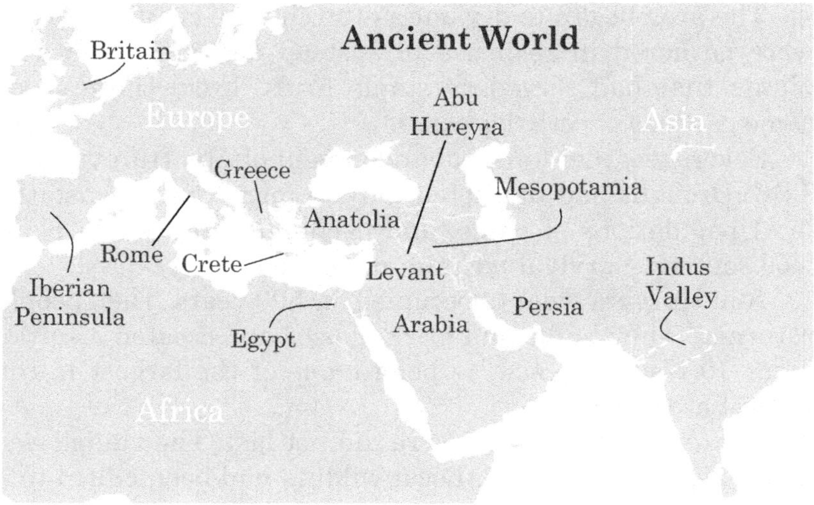

Prior to the advent of agriculture, early villages died out when they exhausted local resources. The resettlement cycle repeated until crop cultivation and trade rendered location itself a significant resource.

Crop cultivation only began 23,000 years before the fall. At that time, women were learning the rudiments of agriculture: cultivating choice seeds and keeping out inedible weeds.

A pronounced period of sporadic but extensive glaciation relented 14 TBF. The settled site of Abu Hureyra illustrated the difficulties after the chill.

Abu Hureyra was in the Euphrates valley, in Mesopotamia. Abu Hureyra was first occupied 13 TBF. It was a village of small huts, cut into the soft sandstone of the terrace there.

The huts were supported by wooden posts and roofed with reeds and brushwood. There were underground food storage compartments.

In its 1st incarnation, Abu Hureyra was a village of a few hundred at most. Villagers gathered wild plants, ate local game, fished. Gazelle migrated through during the summer. Year-round were sheep, aurochs, and onager, a few of which were kept as livestock. Smaller food animals included birds, fox, and hare.

Many early settlements rejected domestic livestock in favor of wild game. Hunting versus herding, and farming versus foraging, were social decisions that varied by location and culture.

The area began to dry out as the climate changed. Seeds were gathered in response to a steep decline in the wild plants that had served as staple foods. From these seeds came gardens of cereal grasses.

Severe weather forced abandonment of Abu Hureyra 10.2 TBF. Drought had disrupted gazelle migration, devastated local populations of game, and decimated forageable plant food sources. Survival meant a more mobile existence.

Abu Hureyra went unoccupied for 500 years. Then people returned. They built mud-brick houses and created a settlement 10 times the size as before: one of the largest in the Levant at the time.

Settlement at Abu Hureyra did not last. The village was again abandoned 7 TBF. Local wildlife had been culled to a degree that their harvest was insufficient. Growing crops was a hardscrabble lifestyle that residents found unpalatable.

Farming communities became increasingly common in the Fertile Crescent from 12 TBF. East Asia followed shortly thereafter, as did other peoples around the world. In the process, cultures changed to reflect these new lifestyles.

In temperate biomes, animal domestication was often decisive in settling down. Because they could be 'harvested' anytime, livestock were considered a more reliable food source than crops, which were sensitively dependent upon seasonal weather conditions.

Early agriculture was a survival strategy. Animal domestication and cultivating crops were a reaction to overexploitation of local resources by a burgeoning population.

Agriculture engendered settled communities. Strife often grew alongside settlements. Settling intensified conflict between neighboring groups for the best land and convenient access to water. Disputes were especially acute when harvests were paltry.

In birthing towns, people wrought exclusion zones from Nature. Not only were wild animals not tolerated, but local vegetation was also much modified. Wastes accumulated in and around settlements, further degrading the habitat.

Erosion from deforestation and intensive agriculture damaged the soil. Growing enough food grew problematic. As at Abu Hureyra, settlements were abandoned when nearby resources were exhausted. This repeatedly occurred in once-fertile places.

Settlements birthed a sense of entitlement to private property. The notion was reinforced by agriculture. For some, surpluses accumulated to produce wealth, which was furthered by inheritance of land.

⌑ ✧ ⌑

The animal that had the most impact on men was the horse. In their descent, horses adapted to survive on steppes and in savanna with sparse vegetation. These were forbidding habitats for other large grazing animals, especially ruminants, such as bovine (cattle), deer, and sheep.

Horses died out in the Americas in the last ice age. But equids endured in Eurasia. Eurasia's larger landmass afforded more warmth near its center, where horses survived.

Horses were sensitive, gregarious herd animals. They were hunted by men, and first domesticated for their meat. Young horses became pets, cementing a social bond.

The first human cultures to ride horses were in central north Asia, 6,000 years before the fall. Horse-drawn travois, wagons, and chariots followed.

The utility of horses found favor throughout Eurasia, where they spread in succeeding millennia. Their introduction often came by force, as an instrument of conquest.

Within a thousand years of first being ridden and driven, horses went to war. Besides tilling fields and for transport, horses were accomplices to the brutality of men for many millennia. Horses changed the equation of conflict in a way that was only superseded by gunpowder weaponry.

Temujin was born on the steppes of Mongolia in 938 BF to a family of nomadic hunters. Severe adversity at a tender age both toughened and seasoned Temujin with forbearance.

In 894 BF, a tribal assembly confirmed Temujin as Genghis Kahn: commander of all Mongolia. Temujin assimilated conquered people rather than follow the steppe custom of looting and leaving them. This sensibility abetted the Kahn in creating a dynasty that conquered its way to the largest contiguous land empire in history.

The logistical lynchpin of the Mongol conquests was, of course, the horse. Helping the rise of the Mongols was warm, wet weather that allowed favorable conditions for breeding horses. Most calvary campaigns took place during the warmer seasons, when there was sufficient grazing for their herds. The Mongols disseminated horses throughout Eurasia, accelerating the animal's cultural adoption.

The cost of 'building' the Mongol empire was tremendous. The Mongol conquests were one of the deadliest in history.

The world before the Mongols went on their bender had 360 million people. Some 50 million were killed during their subjugation. This death count does not include those who died from the bubonic plague, which the Mongols carried

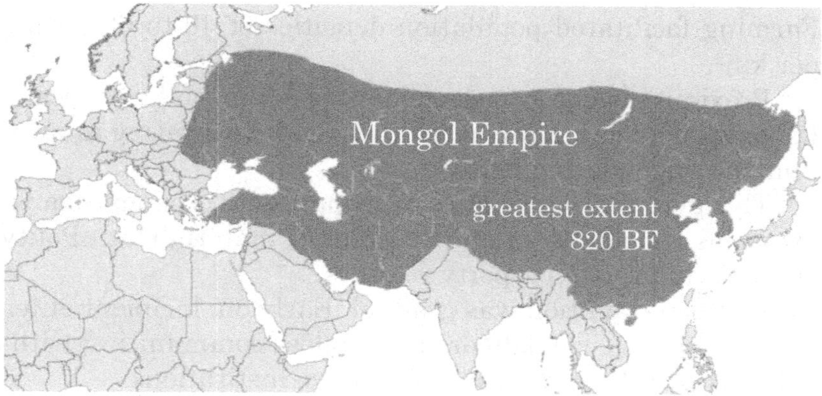


with them. The Black Death that decimated Europe in the 8th century BF was sparked by the Mongol conquests.

The Mongols did not have an overarching concept for ruling a settled population. The Mongol empire was not a state in the normal sense of the word: merely a vast agglomeration of diverse tribes held together by military domination. The Mongols relished conquering, not ruling. Battle was a thrill, governance a tedium.

Already in Genghis's lifetime, the empire was divided among his 4 favorite sons into *ulus*: a Mongol word which denoted supremacy over certain tribes rather than a defined territory.

As a divisible dynasty, the Mongol empire was not built to last. Lacking a strong political or cultural backbone, the initial policy of assimilation transformed into dissolution.

Most conquered tribes reclaimed their lands within a century of their subjugation. Though some political relics remained, within 3 centuries of the first Mongol conquests the last Mongol ruler had been overthrown.

War horses were only retired for motorized tanks after the 1st World War. This was part of a trend in abandoning equids for a dirtier substitute. The rapid replacement of horses with petrol vehicles in the early 2nd century BF was a critical turning point in man's drive to self-extinction.

¤ ✧ ¤

The social impact of agriculture was enormous. Depending upon locale, especially its climate, the population density of foraging societies ranged from 0.1 to 1 person per km$^2$.

Farming facilitated population densities of 40 to 60 people per $km^2$.

Proximity begat interdependence on a much larger scale. Labor specialization furthered the trend, facilitating innovation and engendering trade.

This gyre geared societies to grapple with economic transgressions, molding morality around materialism. Polity evolved via private property regimes.

Political evolution was gradual. Early on, incipient elites lived among their neighbors in stateless communities, without the control mechanisms of coercive institutions.

This made the well-to-do vulnerable. Men garnered power by hiring protection. Authority as a creature comfort had such an appeal that men's appetites for status and wealth grew.

Political states arose first in areas which grew grains as a major crop. "It was not an increase in food production that led to states, but rather the transition to reliance on appropriable cereal grains that facilitated taxation by the emerging elite. When it became possible to appropriate crops, a taxing elite emerged, and this led to the state," reported 1st-century-BF historian Joram Mayshar.

Authorities cultivated cultures that shifted morality from equality to inequity as a norm. There were twin forks to this corruption: material and spiritual.

Persuasive men were able to accumulate property and power. Power meant the resources to dispense livelihoods to others. Work became employment.

Clans evolved a strong corporate identity, built around genealogical connection, whether actual or mythical. The idea of social identity arose. Thus the seed of tribality was planted.

Earthers had an innate sense of spirituality. This grew from the desire to find meaning and purpose in life.

Information fed material, social, and spiritual needs and wants. In an increasingly interdependent environment, control of information meant power.

Authority grew as chiefs gathered skills about them: craftsman to make valued goods, artisans to elevate esteem via aesthetics, priests to weave stories and rituals which sated spiritual wants and sewed social cohesion.

The seminal moral shift underlying the emergence of political states was social acceptance of inequity: a perversion which warped sense of fairness.

Organisms use their sense of equity to assess the quality of their interactions with others. Even single-celled bacteria have a feel for fairness.

Plants regulate their distributions to mutualist microbes based upon the little ones' relative contributions. Social animals, hominids included, were no different.

Empathy is the tender projection of how another feels. Compassion adds incentive to do something when sensing suffering. Lacing empathy with fairness composed morality. 3rd-century-BF German philosopher Arthur Schopenhauer considered compassion "the basis of morality."

Morality was fundamental to how people defined themselves and adjudged others. "Moral character information determines the impressions people form of individuals," noted 2nd-century-BF psychologist Geoffrey Goodwin.

Moral impressions cast the foundation of relationships. "Moral traits are a reliable predictor for how individuals will fare as potential partners for cooperation and affiliation," observed 1st-century-BF psychologist Nina Strohminger.

Mentalizing (mind perception) is the process of inferring someone else's perspective. "Mind perception is the essence of moral judgment," remarked 2nd-century-BF psychologist Kurt Gray. In assessing the morality of acts, people tried to discern intention. "Determining intentionality is the first step in moral computations," wrote 1st-century-BF psychologist Jean Decety.

"Human morality grew out of our primate psychology," observed 2nd-century-BF ethologist Frans de Waal. Simian sense of fairness was selfsame.

What differed among simian species was social tolerance for inequity. Many were even-handed in empathic relations, with insistence on equity at the risk of ostracization. This was consistent with many other mammals and birds. It also corresponded with early hominid morality.

By contrast, those primates with brusque, hierarchical social systems, such as baboons, endured considerable inequality. Like baboons, humans evolved to inequity as an acceptable norm. The moral decay came with economics.

With the advent of private property, sense of fairness in men morphed from the age when foragers shared and held equality as an abiding moral value. "Views on morality changed dramatically across history," observed 1st-century-BF psychologist Lisa Cohen.

Inequality took root with agriculture: once land was given intrinsic value. Agrarian inequity was initially accepted because its origination was ostensibly of good fortune in plot selection and hard work. Farming sharpened ownership and the work ethic into moral values unto themselves.

¤✧¤

Economics was born when there was something to trade. Surplus birthed economic activity.

Economics began innocently. Someone might come to possess something or make more than they needed. As well, they may want for what they don't have. The twinning of excess and want were the wellspring of trade. Settlements were often sustainable only because of trade.

The economic term *goods* dates to 8th-century BF English: something good made by someone's labor. A good kept by its maker was simply a possession. Goods became economic when they became available to others.

Goods had a life cycle. They were created, distributed in some way, and then consumed. Any leftover was waste.

The problem with goods was *bads*: untoward externalities. An *externality* was an unintended effect. Chopping down a tree for fire, to build a home, or clear land for planting crops negatively affected much local life, beginning with the tree.

Bads were the evil twin of goods. What changed with technology was the scope of bads, and thereby their habitat impact.

Along with population growth expanding the quantities of bads, bads from goods accelerated with industrialization. Industrial products were often of unnatural materials and chemicals that did not decompose and were toxic. Enduring pollution was a common externality of an industrial good's lifecycle.

The term *political economy* arose in the 4th century BF for the study of trade in light of the state's role. As an academic discipline, political economy was an outgrowth of moral philosophy. In the late 3rd century BF, the term *economics* gradually supplanted political economy in usage. But *political economy* is apter, as it emphasized that goods and bads were always under the auspices of the authorities.

Like other apes, both genders of hominids had their own dominance hierarchies. Male or female, once an ape had dominance it exercised *authority*: the ability to settle conflicts and set rules. Apes acknowledged authority through submissive greetings and gestures.

Readiness to resort to violence was the constant backdrop of human politics. "Blood alone moves the wheels of history," wrote German friar Martin Luther. An outraged Luther sparked the Protestant religious revolt against the corrupt and domineering Catholic Church in the early 6th century BF.

The availability of abundant resources lessened aggression in other primates. Men were exceptional in this regard. In them, abundance generated greed.

The history of man was a chronicle of coerced inequity in resource availability: a primal economic violence. The beating heart of Earther political economy was coercion enforcing unfairness. "Under capitalism, man exploits man. Under communism, it's just the opposite," quipped 2nd-century-BF economist John Kenneth Galbraith.

Stratification deepened as agrarian societies grew. Rather than communal plots, clans worked fields separately, cooperating only when necessary. Territoriality strengthened the grip of private property and elevated materialism as a moral value.

Communalism versus individualism was a distinction between foragers and agrarians. Still, agriculture involved a sliding scale of social endeavor. The degree of interdependence among farmers depended upon the dominant cereal crop.

A collective culture grew where rice was the primary grain. Before mechanization, growing rice took twice as many hours of work as wheat. To deploy labor effectively, especially during planting and harvesting, rice-growing societies developed cooperative labor exchanges.

The history of social and political dynamics in southeast Asia and Japan, where rice was the staple cereal, testified to the political inclination toward consensus born from collective work in the fields.

Cultures which relied upon wheat and other grains that could be independently grown led to individualism. This was clear in the Fertile Crescent, where concocted religious creeds became the brittle glue of association among its flinty peoples, who were more given to conflict than cooperation.

That privatism would triumph over communism with the advent of agriculture was not inevitable. It was instead cultural.

Among certain peoples, land was communal. The soil was tilled and harvested for all. This was true for the first peoples of the Americas, among some groups in Africa and the Indian subcontinent, and throughout many of the islands south and east of the Asian mainland. Land, like water and wind, could not be sold.

The communist spirit was eventually overtook. Native cultures which practiced communalism were technologically content, and thereby ill-prepared for those men, hardened by individualism, who came with conquest and plunder on their minds.

¤✧¤

The production of surpluses begat a gyre of materialism that defined economics and shifted men's morality. Once the inherent inequality of private property developed as an acceptable norm, it had its own momentum. The idea that Nature may be 'owned' was the crucial concept that set the vector of human civilization.

Consider mode of survival as a risk mitigation strategy. For nomads, sharing is insurance. Foraging and hunting are chancy endeavors, so a regime of sharing is adaptive.

Landing in a rich land changes the risk equation. If abundance is at hand, best to stake a claim to it. Being able to store food and valued goods further damned the sensibility of sharing.

Herding and raising crops are laborious. Only property rights render such a life strategy sensible.

⌑ ✧ ⌑

From antiquity there was a crucial lubricant which engendered trade: money. Money abstracted economic value into symbolic form.

Money originated to overcome the logistical difficulties of barter. Whereas *money* is a generic fungible (medium of exchange), *currency* is a monetary species. USA dollars and European euros were exemplary currencies in the last age.

Currency brings money to life. The term *currency* derived from *currentia* in medieval Latin, meaning "in circulation." Currency was a commodity of social fact: an agreed-upon token signifying some designated value.

Currencies descended from commodities in common use or held in especial esteem. Obsidian was a shiny example.

Obsidian – volcanic glass from Anatolia – was the material of choice for Stone Age tools. For over 8,000 years, starting around 15 TBF, obsidian served as a regional currency. Çatalhöyük and other ancient Neolithic villages in Anatolia were partly supported by having a supply of obsidian nearby. Obsidian as currency was contemporaneous with grain and livestock as units of trade.

Ancient Indians used almonds as specie; southeast Asians, rice; Mongolians, bricks of tea; Guatemalans, corn; Babylonians, barley; Nicobar Island natives, coconuts; Norwegians, butter and dried cod.

Taking a step into monetary abstraction, Mesopotamia in the early 6th millennium used weighed amounts of silver for payments. Silver superseded grain because it did not rot. Centuries later, Egyptians added luster to the idea by using gold.

Early codes of law formalized the role of money in civil society. Among them was the Mesopotamian Code of Ur-Nammu, which wishfully decreed in its prologue: "equity in the nation." Mesopotamians had a vibrant economy via currency. A land of frequent conflict and slavery, equity was in short supply.

The difficulty with currency came in its preservation of value over an expanse. To that end, coins had to have intrinsic value. That meant coins were precious metals: preferably gold, or, of lesser regard, silver.

The use of precious-metal coinage did not obviate other commodity currencies. In Roman times, salt served as a common currency. Roman soldiers were sometimes paid in salt: *salarium* in Latin, which formed the modern word *salary*.

Gold was always considered the most precious metal. Its reddish-yellow color and soft luster had an undeniable appeal. But there was more to gold's allure.

As a storehouse of value, gold had 2 advantages: it was rare and practically incorruptible. No chemical agent known to the ancients could diminish it.

Gold was readily processed and its quality easily assayed.

Gold has a low melting point: 1000°C, almost the same as copper. Pure gold separates from whatever dross it is with by melting it.

Quality assessment is even simpler. The purity of a gold object can be ascertained by putting acid on it. If it does not decay, it is pure gold.

Silver was the other precious metal serving as currency. Silver was much more abundant and not nearly as incorruptible. Significant silver finds at various times in history caused huge price fluctuations that made silver less desirable than gold as a currency unit. As a backing for currency, stability is prized above all else.

Gold is not just rare on Earth. Gold is scarce throughout the cosmos. Iron is the densest possible production by a star before it explodes in a supernova. Gold can only be made colliding neutron stars, which are the densest stellar remnants.

China was the first purveyor of state banknotes. By 3400 BF, notebook-sized sheets of mulberry bark circulated as currency. The sheets carried the emperor's seal and the signatures of the country's treasurers.

China also invented paper money, which evolved in the 14th century BF from deposit receipts, used by merchants and wholesalers who sought to avoid carrying bulky coinage for costly commercial transactions. The Sung Dynasty issued the first government-backed paper money in the 11th century BF.

The notion of paper currency came to Europe in the 9th century BF, through accounts of travelers to the Far East.

As trade expanded in Europe toward the end of the Middle Ages, bills of exchange became prevalent. They were primarily a safety measure. Traveling with cash was dangerous. A deposit could be made with a banker in one town for a bill of exchange, which could be redeemed in another town. Such bills could also be used as payment.

This early form of credit became a significant source for creating money, in similar fashion to ancient Egyptian grain banks, which issued notes from deposits. Bills of exchange were common in England before credit lines became widely available in the mid-3rd century BF.

Coins had been used in England since the 24th century BF. From the 8th century BF, Britain's authorities regulated the precious-metal trade to ensure quality. Goldsmiths were incorporated into local guilds.

Goldsmiths were not merely bullion merchants. They were also money lenders. By creating a temporal rift between assets and liabilities, the very process of lending fabricates money.

The Royal Mint of England became a main repository for merchants to deposit their precious-metal wealth. In 460 BF, King Charles 1 – a staunch believer in the divine right of kings – took advantage of that trust. He seized the private gold stored in the Mint as a forced loan. Thereafter, merchants deemed their gold more secure in the private vaults of local goldsmiths.

This escapade proved fortuitous. Charles' regal caprice led to goldsmiths becoming the forerunners of Britain's modern banking system. They created the instruments necessary to finance investments on a large scale.

Owing to the extent and sophistication of its banking sector, England was uniquely poised to privately finance industrialization when the machine age began, in the 4th century BF.

Paper money had an inherent dilemma: trust in its issuer. As governments took on the monopoly of printing money, confidence in currency became a matter of state.

Governments established their fiscal trustworthiness by ostensibly backing their paper currencies with precious metals. The economic integration among countries which began in the 3rd century BF owed to having an international gold standard. Alas, that metallic metric did not stabilize currencies.

However sensible it may have seemed at the time, precious metal as a salve for states' mutual mistrust turned out to be an unsound insecurity. Then again, irrespective of monetary scheme, stability in fungibility eluded modern Earthers.

Men weaned themselves off metal-backed currencies 130 years before the fall. They did so after an unceasing series of monetary crises, in the wake of wars and spectacular financial speculations that ended in busts.

In abandoning metals as monetary signifiers, governments simply declared what their currency was worth. The reputations of states resolved through trading, resulting in currencies' relative values.

Fiat money became universal out of suspicion: a financial irony in the extreme. Currencies became a national confidence game. This enshrined volatility.

In a sustained fantasy of optimism, over 2 dozen nations in the European Union (EU) forged a common currency – the *euro* – at the end of the 2nd century BF. The euro was a play of plutocracy. While the rich nations in the club made out well, the euro sapped the weaker ones. The weak went along only because they were economic dependents to the richer states.

A currency was not just the coin of the realm. It was also the means to regulate the financial flows which lubricated a society's economy.

The inability of weak EU nations to adjust economic dynamics through financial levers meant their populations suffered a lower standard of living while providing goods and services to their wealthier European cousins at a rigged discount. The imbalance resulted in a series of financial crises which were subdued by submersion. The rich EU nations buried the problem with cheap loans to the poor ones: loans which were perpetually renewed. This was done to sustain the political confederation of the European Union and keep the rigged game going.

The euro was instrumental in the destabilization of Europe as its civilization collapsed, a half-century before the fall. Sapped by the rigged currency, people in chronically weak countries revolted as their cost of living spiraled. This chaos caused the breakup of the EU. That tumultuous experiment in supernational government lasted less than a century.

4 centuries before the fall, germinal economist Adam Smith foresaw the problem with a financial system founded on the quicksand of abstraction: "The trouble with paper money is that it rewards the minority that can manipulate money and makes fools of the generation that has worked and saved." In the last 2 centuries of man, financial manipulation was the main way fortunes were made – and the way that those who had prudently saved saw their financial nest eggs evaporate.

⌑✧⌑

An *economic system* is a characterization of how a political economy works. Economic systems pertain 3 issues: what goods are made and how many, how goods are made, and how goods are distributed. There are 3 economic systems: traditional, command, and market.

A *traditional economy* relied upon custom to address what was made and how it was shared. Traditional economies produced to survive. There was minimal trade. Societies which relied upon economic surpluses to thrive extinguished traditional economies.

In a *command economy*, the authorities decided production and distribution. Command economies were typical in ancient civilizations, including China and Egypt.

In the modern era, command economies were used by Russia, Japan, and China to industrialize rapidly, to establish themselves as a global political power.

Russia was ruled autocratically throughout its history. Holding rigged elections from the late 110s BF did not change that historical constant.

Long an agrarian nation, Russia was a latecomer to industrialization, which it only began in the 220s BF. After a single decade, this modernization was interrupted for 3 decades by economic depression and political revolution.

Joseph Stalin became dictator in 178 BF. He ruled with an iron fist until his death 3 decades later. Stalin imposed a brutal command economy which furthered industrialization at the expense of staggering pollution and 20 million Russian deaths.

Successors to Stalin only did better by not slaughtering so many Russians. Russia largely abandoned its command economy in 109 BF, when its government collapsed in fatigue. The Russian political economy became an oligarchy of state corporatism: plutocratic capitalism via state puppeteering.

After a century of civil war in the 6th century BF, Japan saw stability for 3 centuries with its feudal political economy. Japan's idle of isolation was broken by the threat of invasion in 247 BF by nouveau imperialist USA.

A prodded Japan rushed to industrialize via a command economy. The nation marched under the slogan: "Enrich the country, strengthen the military." The Westernizing Meiji Restoration transformed Japan into an aspiring imperialist power, beginning in the 220s BF. This set the island nation on a collision course with its provocative nemesis: the USA.

The revolutionary who started China's last dynasty was Mao Zedong. With help from the Russians, Mao installed a command economy upon taking power in 151 BF. "This led to the growth of bureaucracy, new social inequalities and privileged elites," wrote 2nd-century-BF China historian Maurice Meisner. The outcomes of Mao's "Great Leap Forward" were societal discord, 30 million Chinese dead from famine, and a billowing beginning of horrific pollution from pell-mell industrialization.

After Mao, China dabbled in state-sponsored corporatism. The unsurprising result was burgeoning inequality.

A *market economy* relied upon disorganized endeavor, using trade as a regulator. Aside from a few ill-fated attempts, the modern global economy was carpeted with capitalism.

Finance propelled modern capitalism, successfully concentrating wealth. In 80 BF, the richest 1% of households controlled 44% of the world's capital while the bottom 50% held only 1%.

The term *capitalism* was coined by socialists as a pejorative in the mid-3rd century BF. Its apt jab was that wealth (capital) was the controlling power behind a market economy. Capitalism crafted this character during nascent industrialization, which sprang from speculative investments that ushered Earther's last age.

As a political economy, capitalism weaponized in private hands the plutocracy that had prevailed throughout history. The term *plutocracy* descended directly from ancient Greek for "the power of wealth."

Market economies produced prodigious externalities in the machine age: inefficiencies, overproduction, unemployment, inequality, corruption, and pollution. As the world then was one disjointed market economy, the fruits of civilization owed to the economic disorder that was capitalism.

While the market system mixed bads with goods, its psychological appeal to ambitious men cannot be overstated. Cast as an ideal of private enterprise, capitalism practically sold itself. Sewn from a sense of entitlement, individualism was a mindset not easily set aside.

By contrast, socialism was a tough sell. Community spirit did not find fertile soil in a ground sown with self-interest.

The materialist mindset increasingly took hold among common folk from the 9th century BF. By the 2nd century BF, creature comforts were the common aspiration. This transition transpired as a growing percentage of a population went from rural living to an urban lifestyle which emanated from manufacture and trade.

The term *bourgeoisie* was first used in France in the 10th century BF for the occupants of walled cities (boroughs). The bourgeoisie occupied the middle level of the medieval caste system, sandwiched between poor peasants and the wealthy. At the societal apex were royalty and landed aristocracy. As economies were enlivened by trade, the *middle class* emerged from this sandwich 2 centuries later.

From the 3rd century BF, especially in industrial nations, capitalism was pitched as a karmic system: by presumed dint of diligence, the rich were winners, the poor lazy losers.

Propped up by corporate advertising, public media companies touted the glamour of wealth. The rich were celebrities when they chose to be. The gullible Collective ate it up. The idiom "money talks" became a cliché. Plutocracy was a cultural embodiment as well as a political-economic reality.

⌑ ✧ ⌑

Bolstered by authorities and the religions they promoted, inequities became socially acceptable. Unfairness was even argued as being aligned with the laws of Nature. 26th-century BF philosopher Aristotle spoke to a common sentiment when he wrote, "From the hour of their birth, some are marked out for subjection, others for rule." Conservative traditionalists held this opinion throughout history, especially those in power, who succored the status quo to preserve their power.

For most of history among sedentary peoples in the world, slavery was the dominant form of labor. Originally, slaves were the captives of a conquered tribe. But once a society established a slave population, the economic benefits of slavery rendered it an abiding institution. The children of slaves inherited their parents' plight.

Even among native peoples, such as in medieval Europe, most were tied for life to the land they were born on. From a population perspective, kingdoms were really serfdoms.

Though urbanization from burgeoning trade and the advent of industry in the late Middle Ages loosened economic shackles, most folk still toiled in agriculture on their home ground.

For over a millennium, the Mediterranean region was cohabited by Christians and Muslims. These religious tribes fought incessantly, capturing and enslaving one another. The religious conflicts of the Middle Ages were a prototype for the wars of political ideology that came to fore in the 2nd century BF.

During Europe's Age of Imperialism, slaves became a thriving commodity. Indelibly marked by their dark skin, Africans were the favored enslaved. Sub-Saharan Africans themselves had a tradition of slavery among their contentious tribes.

Only toward the end of the 4th century BF did slavery develop into a moral issue: and then only in the richest nations, as its economic import peaked.

Slavery was an institution that Earthers ostensibly outlawed but never did abandon. In the early 1st century BF, some 100 million were enslaved, including children. Over 5 million girls and women were sex slaves. Showing their immoral essence, states worldwide used those imprisoned as slave labor.

Though supposedly given "equal opportunity," people of color who descended from slaves in the USA and Britain were subjected to relentless discrimination in every aspect of their lives, from education to employment to justice. Similar subjection of subordinates was practiced worldwide.

¤✧¤

By the modern age, the socialism of hunter-gathers had become a utopian fantasy. In his day, socialist economist Karl Marx was a wild-eyed radical in proposing that equity was not only technologically possible but morally desirable.

In the mid-3rd century BF, London, England was the epicenter as industrialization began its inexorable grind. Marx was there, soaking it in. Marx observed, "Capitalist production develops technology, combining together various processes into a social whole, only by sapping the original sources of wealth: the soil and the labourer. There must be something rotten in the very core of a social system which increases wealth without diminishing misery."

The consequences of industrial capitalism fill the following chapters of this book. A key question is why the harms went unchecked. That moderns tolerated gross inequity, even when they had the power (via democracy) to bury it, is a key piece of the puzzle.

Over millennia the Collective became inured to inequity. The core tenet of morality was weakened in many minds, but not forgotten by all. "Our world hinges on moral foundations," righteously taught 2nd-century-BF American civil rights advocate Martin Luther King Junior: an eloquent black man who staunchly advocated equity, and who was assassinated in 132 BF by an agent of the USA state for that advocacy.

That capitalism was never abandoned, nor plutocracy overthrown, can be partly explained by the Collective take on ethics, which evolved to a muddle. This was most apparent in legal principles.

The human ideal of justice was grounded in morality. Restitution was the abiding principle of justice systems.

*Ethics* is the philosophy of morality. Humans had 2 contrasting perspectives on ethics: *deontology* and *teleology*. Teleology was also known as *consequentialism*.

Under deontology, morality is in the act, not its fallout. Deontology is morality by rule, not result.

The refined distillation of deontology is *moral absolutism*: that certain acts are morally right or wrong regardless of intent or consequence. Few people were moral absolutists.

Consequentialism eschewed intention as the pointer in the moral compass. Teleology instead cast morality in the outcome of an act.

In determining crime, modern justice systems capriciously mixed deontology, teleology, and psychology. The mixture depended upon the violation.

Of highest priority to authorities was respect for their social stature. Any expression of ill intent against the government was considered criminal.

Only when it came to submitting to the state did moral absolutism fit the bill. Deciding other crimes depended upon consequence or intent.

For a party determined as responsible, carelessness (negligence) might be a violation when there was a consequence to it. The outcome of an act was often instrumental in deciding the severity of punishment for a crime.

Depending on the deed, there was more than consequence to crime. Most often, what determined whether a crime or accident occurred was the perpetrator's intent in performing an act that had a criminal outcome.

Psychology defined morality. Coincidental causality did not trigger outrage, however systemic it was. Intent defined the ethical essence of a situation: what was the perpetrator *wanting* to do?

In considering the implications of acts, irrespective of intent, moral absolutism would have recognized the destructive exploitation of capitalism for what it was. But men were not naturally moral absolutists.

Conversely, a purely teleological view of capitalism would also have earned that system moral condemnation. But consequence was considered incidental to intent.

Trade by barter is an apples-and-oranges exchange. Values are subjective, based upon respective wants. Morality did not weigh in with such a trade.

This calculus continued into the modern age. Individual mercantile transactions failed to give an impression of unfairness, especially because currency inserted a level of abstraction that blurred any inherent inequity. Obvious in its exploitative intent, price-gouging was a notable exception.

The capitalist system was not viewed immoral because it ostensibly lacked motive beyond an agreed exchange. Failure to find malice fizzled feelings of wrongdoing. Drained of emotion, indignation had no legs.

For moderns, the inequality of the market system offered opportunity as well as oppression. Capitalism's pervasive inequity only invoked widespread moral condemnation when its intrinsic deficiencies were sorely exposed, such as during the prolonged Great Depression in the 160s BF.

The inevitable bads produced by an economic system geared to ruthless exploitation and inequity were commonly dismissed by capitalists as fixable spot flaws. But these defects never were remedied. Nor could they be. The externalities – moral debasement, inequity, ruined lives, waste, and pollution – were as inherent to capitalism as exploitation, which was the spark of intention in the economic engine.

Only in the last century, after self-extinction was already assured, did a woke minority come to consider the desecration of Nature a crime. It was hard for an omnivore which lived off other life to make a moral infraction out of killing non-human animals and plants.

Further, condemning the economic system which naturally perpetrated this crime was an abstraction of magnitude few grasped. The idea of overthrowing the global economic system to 'save the planet' seemed a hopeless case – which it was.

More than any other mammal, humans socially evolved into interdependence. Culture acted as social glue. Societies thrived or faltered owing to their core cultural character, which was expressed in various folkways and pivotal mores.

An entitlement mentality was widespread in the peoples of industrialized nations in the last 2 centuries. This cultural trait came from overlaying the idea of "human rights" onto a materialist mentality. It was a muddled merging of capitalist and democratic values into a banal banner of natural law.

Humanity's mangled sense of morality was instrumental in its demise. Smothering fairness with materialism was never going to end well.

❀❀❀

A *tribe* was a cohesive group sharing similar folkways. Tribes did not arise until settlements were conquered by men who then needed to somehow stitch social bonds. Tribality was both the result of conquest and the progenitor of war. Via territoriality and subjugation, humanity became a world of tribes.

Violent subjection was typically by men who hunted or herded. Their target: crop growers.

Cultivation taught patience and pacific ways. Growing crops inured men to a prosaic routine and exhausted them from a long day's toil. Such men accumulated wealth but lost the craft and sentiments for war.

Hunters and herders were accustomed to danger and skilled in killing. War was but another form of chase and held hardly more peril. When the woods ceased to give abundant game or flocks thinned, the stores of a prosperous village proved too tempting.

"Property was the mother, war the father of the state," observed Will Durant. A *state* was an abiding political institution, attended to by authorities. The state ruled a nation.

A *nation* was a political territory with which residents might have felt identification. Emotional allegiance was termed *nationalism*.

The controlling authority of a state was the *commander*. Whereas a *ruler* had limited power, an *autocrat* (*dictator*) had absolute authority.

Authority was typically weak in preindustrial tribal societies. A tribe leader had limited ability to coerce individuals.

Virtually all tribal societies had traditions for justice. These included sanctioning kinsmen to seek revenge or restitution for wrongs, nonbinding arbitration, and setting restitution schedules. Tribes typically lacked an effective system for 3rd-party rule enforcement.

Blood feuds were a tribal institution. Fear of incurring a blood feud was the most significant legal sanction within a tribe, and the greatest guarantee of life and property. There were rules as to how blood feuds were to be pursued.

Popular assemblies originated as a venue to adjudicate disputes. The *Illiad*, a 14th-century-BF Greek epic, described a feud over the blood price for a slain man, argued before a marketplace crowd. The verdict was announced by tribal elders.

Tribal societies did develop increasingly sophisticated institutions for civil and criminal disputes. But 3rd-party enforcement of those decisions only came with the emergence of states. It was in those early states that laws evolved from moral traditions, often cast into religious dogma by way of creative scribes penning mythical gods.

The state was a late development, hardly appearing before men took to writing history. The emergence of the state was a fundamental change in social order, from family ties to domination. "The state as distinct from tribal organization began with the conquest of one race by another," wrote 2nd-century-BF historian Lester Ward.

The state fortified itself for war and evolved in the aftermath of conquest. Only conflict created the conditions by which men would bind themselves together regardless of kinship. States arose as a vehicle for allocating goods among favored dominants, and for sustaining subjection of subordinate groups, especially those considered troublesome.

The emergence of states did not solve the problem of violence: quite the contrary. Statehood instead institutionalized violence. "The state is a product of force, and exists by force," observed sociologist William Graham Sumner in the 2nd century BF.

Governance was most important when tribes were conquered and survivors absorbed. Rules of conduct were especially needed in reducing conflicts among groups with their own distinct culture and norms. Government cut its teeth on post-conflict social management. "Under the tutelage of the state, human beings learned for the first time how to bow, grovel, kneel, and kowtow. In many ways, the rise of the state was the descent of the world from freedom to slavery," explained 1st-century-BF anthropologist Marin Harris.

To ensure stability, the abiding authority of a state could not rest on force alone. To maintain dominion, authorities forged instruments of indoctrination.

Religion was both the most abstract and most common societal glue that the state used. In secular modern states, nationalism substituted for religion.

Democracy provided a periodic vent for citizens, though it never changed how the state operated, or to what purpose: self-preservation of the state. Whatever services they provided, states ultimately served themselves. Only revolution could overthrow that, albeit with an outcome that was always the same: institution of another state with the same purpose: sustaining the state.

To the end of Earther civilization, a feudal system kept social order. States codified codes of conduct and allocations with a body of law. Laws were readily superseded when it served the state.

¤✧¤

Almost all large civilizations originated through conquest. The ancient empires of China, Mesopotamia, and Egypt were hierarchical, with social castes, slavery, and rife with seething class conflict.

An exception arose in the Indus Valley, the fertile floodplain of the Indus River. The Indus civilization was an epitome of human society, and its shortcomings emblematic.

The Indus civilization began 9 TBF. Farming settlements sprang up, primarily based upon wheat and barley. Another key crop was cotton.

Livestock were domesticated. Among them were goats, sheep, cattle, and water buffalo. Dogs were part of Indus households.

Baked bricks were key to the rise of the Indus civilization. Their invention permitted the construction of flood-control works on a scale large enough for vast areas to be brought under cultivation.

Detailed craftwork developed, including a wide variety of tools, pottery, dyed textiles, jewelry, sculpture, and seals. The quality and erudition of Indus statuary was comparable to the works of the classical Greeks many millennia later. A written language evolved, with a script that post-industrial people could not decipher.

Around 5500 BF, the Indus civilization had burgeoned to over 5 million people. They lived in 54 cities, including 2 major metropolises. There were over 1,000 settlements in an area at least 800,000 $km^2$. Buildings were of baked brick.

Vast stone platforms underlie some cities to keep them above the annual floods. Streets were planned, generally oriented along the 4 main points of the compass.

There were sewage systems worthy of the last age, public baths, and toilets. Given the emphasis on sanitation, public health was clearly important to the Indus people.

All evidence indicates competent governance, but not autocracy, as was the norm in other ancient civilizations. Artifacts suggest a general egalitarianism. Burial remains show little material inequality.

What was perhaps most surprising about the Indus Valley civilization was what was not there. No grand temples. No royal palaces. No slavery. And, judging by the absence of weapons and fortifications, no conflict.

The only fortified settlements among Indus dwellers were on the Arabian Sea coast, where trade with foreigners gave unease. Indus artifacts gave "no indication of a society engaged with, or threatened by, war," wrote 2nd-century-BF historian Neil MacGregor.

The Indus Valley people had geographical luck. Natural resources were plentiful, including abundant forests and freshwater.

The Indus civilization began with settlements close to Kotla Dahar, a deep lake. The climate afforded 2 growing seasons, thanks to winter cyclones and summer monsoons.

Over several thousand years, deforestation and resultant climate change led to the decline of Indus society. As with civilizations throughout the arc of Earthers, the Indus felled their forests and degraded the soil that their lives depended upon.

Farmers shifted their crop patterns as soil quality and the monsoons gradually declined from 5000 BF. This resulted in lower agricultural productivity, which acted as a catalyst for de-urbanization.

Dramatic decreases in rainfall, likely caused by deforestation, led to the progressive lowering of Kotla Dahar. By 4000 BF the lake was ephemeral.

The Indus civilization did not collapse. It simply diffused, leaving behind a depleted ecosystem.

Hindu spiritual and cultural traditions began during the Indus civilization. Though florid and often cryptic, writers of that epoch's Vedic texts show an understanding of the immaterial nature of reality (energyism). That this paradigmatic knowledge was widespread was a novel accomplishment among societies.

Sculpture suggests that yoga and meditation were practiced. But there is scant evidence of organized religion. In this, the Indus people were unique. In all of human existence, the Indus civilization may have been the only one engendering enlightenment.

⌑✧⌑

Southern Mesopotamia lay in the delta between the Tigris and Euphrates rivers. It was an odd spot to cradle a civilization. The flat, river-made land had no minerals, no stone. Typical summer temperature was 40°C, often climbing to 50°C. Annual rainfall averaged 150 mm. It was bone dry for 8 months of the year. Winter nights were shivering cold.

Strong north winds delivered squally rainstorms. The melting snows of spring brought flash floods, which in bad years sweep everything away.

But the alluvial soil was rich. There was abundant timber. Nearby was a great primeval forest, stocked with sturdy cedar.

The Sumer civilization began ~7 TBF. The earliest settlers drained the marshes to grow crops.

The key to Sumer civilization came in water delivery. Sumer cities were the first to practice year-round intensive cultivation, with organized irrigation, monocropping, and specialized labor.

The first settlers in Sumer formed small, self-sufficient communities. As rivers changed their beds, crop irrigators were obliged to cooperate. Water systems were linked, thus progressively enlarging settlements.

Control of irrigation in Sumer created labor classes. Those in control of water and land profited from those who were not.

Sumer's social organization was driven by the need for large public works to overcome climactic challenges, and to exploit agricultural potential in an unsustainable way. Like countless succeeding civilizations, Sumer was bound to eventually fail.

Sumer had industries including weaving, leather work, masonry, metal work, and pottery. The plow appeared early, as did the ax. Trade in all manner of goods became an enterprise.

As early as 8 TBF, within 1,000 years of the Fertile Crescent having settled communities, villages were being abandoned. Erosion from deforestation and intensive agriculture had sorely damaged the soil, making it impossible to grow enough food.

The forests there were felled as quickly as they could be. So great was the abundance, deforestation took over 2,000 years.

With the natural bounty unrecoverably stripped, future generations were to inhabit a forbidding place. What was once the Levant was among the first regions where humans went extinct, so great was the heat on the long barren land.

By 5600 BF, Sumer had been carved into a dozen city-states, divided by canals and boundary stones. Each was administered by an autocrat, commanding under the aegis of a patron deity of the city. Religion was ever the handmaiden of politics.

Ur-Nammu founded the 3rd Sumerian dynasty ~4150 BF. The earliest known laws were promulgated in his name.

The prologue to the *Code of Ur-Nammu* stated that the king was the beacon of justice for his land: a role not normally taken by the king. Disputes were handled by local authorities (mayors), and there was an appeal process: to the provincial governor, and possibly the king.

The *Code of Ur-Nammu* was written as a set of conditional statements: "if… then" situations and redresses. Nearly all later law codes followed this pattern.

Ur-Nammu codes were more humane than the talionic ("eye for an eye") principle behind later Babylonian law. There were monetary compensations for minor injuries and injustices. That withstanding, murder, robbery, rape, and adultery were capital offenses.

After the demise of Sumerian dynasty, from 3994 BF, the short-lived empire of Babylonia arose by way of conquest. The empire emerged from a small kingdom in the minor town of Babylon.

The 6th Babylonian king, Hammurabi, ruled from 3892 to 3850 BF. Revered as a great conqueror, Hammurabi was seen by many as a god in his own lifetime.

His laws outlived the empire. The *Code of Hammurabi* influenced Roman and subsequent law.

By the time of Hammurabi, almost all traces of tribal custom had disappeared from the law. Hammurabi's code served the state's interest in sustaining its authority.

"Hammurabi's code proclaimed that one of its chief goals was to defend the weak against the strong. Yet in early civilizations there was no concept of equality before the law. On the contrary, social inequality was expressed and defended in many ways in legal systems. Crimes committed against the upper class tended to be punished more severely than crimes against commoners. The privileges of the upper classes were protected by armed forces and the legal system," wrote 2nd-century-BF historian Bruce Trigger.

Babylonian law was primitive in its physically abusive punishments for crimes, which were graded depending upon severity of offense and social status, most starkly whether slave or free man.

Nearly half of the 282 Hammurabi laws dealt with commercial matters. 1/3rd of the codes were family law or regulated sexual conduct. A few provisions related to military service. Only 1 law imposed any obligation on an official: that a judge who reached a wrong decision be fined and permanently removed from the bench.

Ancient Jews had a different take on legality. Rather than viewing transgressions solely as societal infractions, the Jews considered crimes as offenses against God.

The *Ten Commandments* were the most famous codes of Mosaic Law. These were traditionally believed to have originated by Moses, who lived in the 37th century BF. Moses was a guru: revered in Judaism, Islamism, Christianity, and other Levant religions. Mosaic laws pertained to crimes, morality, family, diet, purity, and rituals. With rules covering food and bodily secretions (menstruation and ejaculation), the Jews were meticulous in codifying their culture of disgust.

Sexually repressive mores and the attendant taboos of the ancient Jews were passed down, and transmitted to gullible gentiles, for endless generations afterwards. The gender and sexual biases of Western civilization originated with ancient Hebrew codes of conduct.

Ancient India and China had their own legal traditions.

In India, the authoritative *Arthashastra*, on ethical societal governance, influenced the legal texts that followed. The precepts of law in ancient Hindu autocracies extended into southeast Asia and Oceania.

Chinese social philosopher Confucius opined in the 27th century BF, "If the people are led by edicts, they will try to avoid punishment, but will have no sense of shame. If led by virtue, they will have a sense of shame and moreover will become good."

Despite Confucius' moral influence, Chinese law initially favored *legalism*, which held that societal harmony cannot be left to philosophic virtue. Authorities went with strong state control and absolute obedience to authority. Despotism was codified in a web of laws.

The brutal implementation of legalism by the short-lived Qin dynasty (2321–2306 BF) led to that dynasty's overthrow.

Legalism lay in disgrace for 25 centuries, until revived by 2nd-century-BF despot Mao Zedong, who rejected Confucianism. "Chinese civilization never evolved a strong sense of either private property or individual legal rights," observed Bruce Trigger.

The ancient Greeks had no overarching concept of law. Instead, they distinguished between natural law (*thémis*), human decree (*nomos*), and custom (*díke*).

Beyond customary laws, Greece's city-states added codes related to civil rights and constitutions of democracy which proved influential through the ages. Civil rights only applied to franchised citizens. Following custom, most of those in ancient Greece were slaves or otherwise subject to the whims of legal caprice.

The Modern World

¤✧¤

In rendering conquest a permanent condition, domination eventually submerged into normalcy. The French who rebelled in 311 BF had forgot that the aristocracy which had ruled them for a thousand years were descendants of German conquerors.

When tribal-level societies evolved into those ruled by a state, tribalism did not disappear; quite the contrary. State institutions were mere veneer over tribal custom. Both existed in an uneasy balance. Even to the end, ethnically

The Modern World

diverse societies functioned according to tribal norms, with laws and institutions favoring the dominant tribe. Political parties were essentially tribal affiliations.

Tribal structures remained the social locus for societies throughout history, and strongly shaped interactions with modern political institutions. Institutions themselves took on a tribal cast.

Many state institutions tasked with regulating aspects of commerce (bads more than goods) became captive creatures of the industry to which they were overlord. This corruption (tolerance of bads) partly explains the inertia in responding to the dynamics of self-extinction.

❀❀❀

Men found solidarity in war. Sharing harrowing experiences forged enduring social bonds. Many men enjoyed conflict for that reason, as well as the opportunity for legitimate cruelty.

Human history was replete with savagery on whatever scale technology afforded. In modern times, hardly a decade passed without a war somewhere.

Aristotle observed that "warlike races are prone to the love of women." Along with resource acquisition and revenge, the carnal urge motivated taking up arms. Having one's way with women was a principal impetus for war in primitive societies.

While men did not generally associate attractive women with aggression against other men, experimentally exposing men to pictures of desirable women increased their professed support for group conflict. Symbolic images, such as national flags, did not elicit the same response.

That is not to say that symbolism was bereft of power to persuade. "Men will kill and die for an idea," noted 1st-century-BF anthropologist Scott Atran.

When it came to war, women stood apart, beginning with near universal ambivalence. With insignificant exceptions, women were not participants in war – at least until the last century, when 'liberation' meant that women might join in a horrid endeavor they had previously been able to avoid.

Men were not ones to forsake grudges. 52 wars lasted a century or more. 23 went on for over 200 years.

The longest war lasted 772 years, from 1380 to 608 BF. Called *Reconquista*, for "reconquest," the protracted contest was for control of the Iberian Peninsula, comprising modern-day Spain and Portugal. Christian forces battled to expel conquering Muslims from Iberia, and eventually succeeded. Propelled by the bile of righteousness, religious wars were often the most ferocious.

⌑✧⌑

Almost all early societies developed a warrior class. Within this elite evolved the concept of *honor*: the willingness to risk one's life for the respect of other warriors.

Historically, professional soldiers were typically loath to trade places with farmers or merchants, for reasons only partly owing to booty. The lives of peasants and traders were mundane compared to the thrill of danger and the comradery of fellow combatants.

The warrior ethos survived into modern times. Coercion, extortion, confiscation, and gratuitous violence were commonplace among policemen worldwide. That noxious litany withstanding, police believed in their honor.

Wars were a mercenary exercise for most of history. Nationalism did not arise until democracy made it a necessity. Nationalism ostensibly held countries together in times of war, though it did nothing to prevent civil wars.

Militarism pervaded governance until the fall. Harsh military training substituted for the rigors of war in instilling commitment to the ideal of serving a "higher" cause: the state. Militarism also offered a reboot switch when elected politicians proved sorely incompetent: a military coup reset the state to its primal sovereignty.

⌑✧⌑

War was contagious. Peaceable societies had conflict inflicted upon them, leaving no choice but to resist or surrender to slavery. This woeful story was countlessly repeated throughout history.

A notable period of conquest was the Age of Imperialism, from the 6th century BF to the 2nd. The initial aim of seafaring Europeans was exploring for exploitable resources. This "Age of Discovery" naturally led to conquering and/or exterminating the autochthones discovered in "new" lands. Europeans called the hemisphere of the Americas the "New World."

Colonialism was a way to keep milking the extra-territorial cow. European powers first rushed to colonize, belatedly joined by the United States of America (USA). In reaction to being threatened by the USA, Japan briefly joined in.

Purely ideological war came to the fore in the 2nd century BF, instigated by states with overweening arrogance and an itch to engage their military over nothing more than political philosophy. Ideological war represented an astonishing abandonment of self-interest for geopolitical jockeying. If booty was the point of war, squandering resources in military adventure over style of governance was lunacy.

There were 2 global conflicts in the 2nd century BF. Europe was the epicenter of both: a continent long riven by petty politics. Germany, a powerhouse since Roman times, lost both wars.

The 1st World War was also called the *Great War*. World War 1 ran from 186 to 182 BF. 40 million died from the conflict. The unparalleled scale and devastation of the 1st World War owed to the weaponry afforded by industrialization, and to military tactics which astonishingly disregarded the wrath of the munitions used. Thanks to advancing technology, the 2nd World War poured on even more firepower.

In its immediate aftermath, contemporary Europeans had figured the Great War "the war to end all wars." It was anything but. Just 2 decades after the 1st World War ended, the 2nd began.

The peace treaty that concluded the 1st World War ensured a 2nd. France's truculence toward Germany resulted in its demand for impossible reparations. Unmolested by the Great War, the USA was the dominant world power. The USA approved an odious peace treaty which guaranteed that Germany would collapse economically, and that the Germans would take revenge as soon as they could.

After an onerous decade of economic malaise, the German economy had begun a tenuous recovery from its terrible depression in the early 170s BF. Meanwhile, the rest of Europe struggled to recover from its protracted self-destruction.

In 171 BF, the global Great Depression struck. The German government descended into chaos.

In 167 BF, Adolf Hitler had maneuvered himself to be handed power as ruler of Germany. Hitler's ascent spoke of a politically driftless people looking for a strong leader to restore their self-confidence.

A potent public speaker, Adolf had risen politically via artful propaganda which portrayed him as a cultured gentleman and a moral man, fond of children and dogs. Hitler deftly aggrandized his grant of power into despotism.

In starting the 2nd World War by conquering neighboring countries, Hitler set the world on fire. The ashes of that war polluted world civilization long after the guns fell silent.

In its fit of militarism, Japan invaded China in 169 BF. In response, USA ruler Franklin D. Roosevelt (FDR) tried to deprive Japan of oil and other raw materials. The USA incited Japan to attack, which Japan duly did in 159 BF.

The USA was diverted by the European war theater when Japan struck. With a sneak attack on the USA's Pacific navy, the Japanese dreamt that they could strike a decisive blow. That did not happen. The USA was a resource-rich country, and, at the time, was manufacturing full throttle.

FDR died in 155 BF. The war in Europe had ended, but the showdown in Japan had not yet happened.

2nd commander Harry Truman succeeded FDR upon his death. Truman ended the 2nd World War in the Pacific by casual slaughter with atomic bombs. "The most terrible bomb in the history of the world," Truman wrote in his diary.

These freshly developed holocaust weapons were dropped on Japan as a munitions experiment. Hundreds of thousands were killed in an instant, and many more crippled for the rest of their short lives. Thus the nuclear age gruesomely began.

World War 2 further darkened the complexion of humanity. And accelerated its demise. Out of that terrible episode came numerous trends which might not have otherwise occurred.

The American atomic bombs dropped on Japan were a prelude to over 2,000 stronger nuclear detonations in the decades that followed. 8 nations took part in gratuitously polluting the planet: USA (over 1,000 "tests"), Russia (over 700), Britain, France, China, India, Pakistan, and North Korea.

These atmospheric bombs destroyed vast swaths of land and spread radioactive waste all over Earth. The worst of these tests, by the Russians, was 3,300 times the "yield" of the atomic bomb which destroyed Hiroshima.

In the aftermath of the 2nd World War lay the last, best opportunity for mankind to reconsider the ways which had been shown as folly. It was an opportunity not taken, as the world's leaders simply did not deserve the offices they had been awarded.

While the rest of the world was (again) exhausted from the (2nd) global war effort, the USA was invigorated. By its victory and absence of war damage, the USA remained the leading world power, both economically and politically. This victory gave the USA a hubris which left in its wake a legacy of failed ideological wars: in Korea, Cuba, Vietnam, Iraq, and Afghanistan.

During the 2nd World War, Dwight Eisenhower had been commander of the prevailing army in Europe (over Germany). He was subsequently USA ruler from 147 to 139 BF. Eisenhower observed, "Every gun that is made, every warship launched, every rocket fired, signifies in the final sense a theft from those who hunger and are not fed, those who are cold and are not clothed." It was lip service. Eisenhower ignored his own advice.

Eisenhower deepened the USA in a losing war against Vietnam for no reason other than not liking the way that nation was governed. America had absolutely nothing to gain, yet committed its dignity, and risked its economy and social fabric, to invade Vietnam. It was a roll of loaded dice that the aggressor was destined to lose.

The USA political establishment at the time had no conception of what was at stake. Ignorant of history and their chosen enemy, the American political elite naïvely assumed they could prevail through destructive firepower, though to what end they apparently had not bothered to ponder. Colonialism was not an option.

Over 3 million perished in America's violation of Vietnam. Whereas the war brought Vietnam together, it divided American society, creating a tribal chasm that lasted a generation.

❀❀❀

Throughout human history, states governed autocratically. Democracy was an outlier which had no significant impact on the way societies were managed beyond generating dissent and dysfunctionality.

The term *democracy* originated in Athens, Greece in the 27th century BF. The wealthy men of Athens and the surrounding territory of Attica decided how to govern their city-state. Meanwhile, slaves, who comprised over half the population, did the work. Athens primarily prospered as a trading center. But traders had no say in how civil affairs were conducted.

The Athenian experiment lasted only 2 centuries before being broken by war. Athenian democracy was in fact brutal plutocracy. "Tyranny arises naturally out of democracy," Plato observed, having seen it firsthand. Athens served as a role model for later democracies.

¤✧¤

Western political regimes were loosely modeled on an ancient civilization that began as a tiny hilltop kingdom, surrounded by marshes, in central Italy. This auspicious start on Palatine Hill was 2853 BF.

Roman autocrats were elected to serve for life by the landholding men there. None in the line of 7 commanders gained or kept power by force. But the last in that line lost power that way.

In 2609 BF, tyrannical Tarquin the Proud was overthrown. The clans that managed this coup grappled to a standstill. The resultant Roman Republic lasted 500 years, until exhausted by civil wars.

In the mid-22nd century BF, general Julius Caesar parlayed his military success in Gaul (modern-day France) into dictatorship. The Roman Empire was born.

Assassinations and other political intrigue saw an unsteady succession of autocrats. The Roman Empire degenerated from the top down.

The overarching structure of Roman society largely followed those that proceeded it since agriculture became a pyramid scheme: a tiny elite lived off the toil of the masses, including an army of slaves which comprised roughly 1/3rd of the population.

The key to Roman longevity was its economics. More than any civilization prior, the Romans esteemed trade.

The legacy of Rome was its legal system. The empire was the first with its law largely oriented toward commerce. Though 2nd-class citizens, women could own property and engage in business.

The premise of Roman law was to create an institutional bulwark which preserved the status quo. Law was a byproduct of a political regime bent to private property rights.

Without claims of property, and attendant laws to protect such claims, there was no soil for crime to sprout. Law created crime.

The edifice of Roman civil law, bequeathed to later civilizations, was plutocracy. Like every society that followed, Rome's justice system was corrupt. The lower classes were fined and imprisoned while the elite were let off.

"In early civilizations, the upper classes and the major institutions they controlled had far more wealth and privileges to protect than did commoners, and they relied on the law to do this. In some early civilizations, growing concern to protect property resulted in increasingly severe penalties for theft. An important function of the legal system was to reinforce the power and privileges of the state and upper classes. Between men of unequal status, the more powerful was invariably deemed to be in the right, and therefore litigation could occur only between equals," wrote Bruce Trigger.

The term *equity* derived from Roman Latin (*aequitas*). From the 480s BF, equity was ironically perverted to mean a right to an economic asset. Nascent capitalists appropriated words to promote this ruthless, exploitative economic system as equitable.

The decline of the Roman Empire occurred over 4 centuries. A catalog of ills took their toll: inept autocracy, economic decay, inflation, sovereign debt, lead in the drinking water (from using lead pipes for public water distribution), social corrosion (slavery, debauchery among the elite, disintegration of army discipline), and constant incursions by unconquered Germans adhering to the Roman pillage model.

Food shortages and the civil unrest it provoked played a role in Rome's collapse. Monocrop wheat grown in distant lands exhausted the soil there. Meanwhile, Italian growers of olives, grapes, and other vegetables prioritized exporting their products for the profits they brought in.

Roman agricultural technology rivaled that of contemporaneous China. It would take a thousand years for any part of Europe to reclaim that level of crop productivity.

Rome finally fell to Germanic invasion in 1624 BF. From then, Germany played a pivotal role in European politics.

A diminished Roman sovereignty held sway in the east for another millennium. This Byzantine Empire eventually fell to invading Turks in 647 BF. Turk conquest engorged the Ottoman Empire, which lasted until 178 BF.

Europe went through a Dark Ages following Rome's fall. The lingering malaise lasted nearly 7 centuries.

Many European cities prospered from the 11th century BF. One the first regions to flourish was northern Italy, as it lay on the trade route between the eastern Mediterranean Sea and northern Europe.

Yet northern Italy was politically insecure: crushed between rival claims of imperial Germany to the north and papal states to the south. Thriving but threatened, the cities sought greater control of their destiny. They adopted the oligarchy of the Roman Republic. Historians called this medieval political regime a *commune*.

Several cities in northern Europe followed the Italian example of communes in succeeding centuries. As in Italy, powerful communes emerged in major trading hubs.

The seed idea of parliament was as an advisory council to an autocrat. Parliaments began in the 10th century BF in the monarchies of western Europe. A parliament was summoned when a despot desired advice.

A crucial step in the evolution of parliaments was the inclusion of wealthy urban burghers (merchants). They were the 3rd estate: an addition to the 2 estates of royal council, which were nobles and bishops. Nobles were the secular aristocracy, bishops the men of power in the (Catholic) church. Spain hosted one of the earliest such parliaments in 912 BF.

In 885 BF, England established the strongest tradition of parliaments. This 1st parliament served as a check on uppity King John, whose reign ended in rebellion.

Parliaments popped up throughout much of Europe by the 6th century BF, varying somewhat in the constitution of their characters. Owing to strong popular opposition to royal pretensions, Sweden's parliament included a parcel of peasant men. This inclusion of crusty commoners started in the mid-7th century BF.

In later centuries, parliaments evolved into legislative bodies and a political check on rulers. In modern democracies, the parliament was one of the tripartite bodies that comprised the core state institutions: commander, parliament, and judicial bench (court).

¤ ◇ ¤

In the 10th century BF, Muhammad al-Idrisi, an Arab geographer, created a world map for King Roger 2 of Sicily. This map, termed *Tabula Rogeriana* ("the map of Roger"), inspired curious explorers throughout Europe. What they lacked were suitable ships and compasses to guide them.

Unsurprisingly, people in Oceania pioneered nautical crafts. Malaysians invented seafaring sails in the 24th century BF. The Javanese had maritime merchant ships by the 21st century BF. Europeans would not learn of these oceangoing technologies for 13 centuries.

The Chinese invented the mariner's compass in the 11th century BF. Europeans did not catch on to compasses until 3 centuries later.

The Chinese were early explorers of nearby lands. The famed Silk Road connected commerce between China and Europe from the 24th century BF. Contemporaneously,

Chinese envoys sailed into the Indian Ocean. In the centuries that followed, China sent fleets throughout southeast Asia, into Oceania, and to Africa.

China's outward push was reversed in the 8th century BF, in the early Ming dynasty. The haijin (sea ban) was kept for 2 centuries, during which the Great Wall of China took its final form. The Great Wall was built to keep pesky foreigners out. Isolationism put China behind in both trade and technology.

Only in the 8th century BF did Europeans gain the technical wherewithal for oceanic exploration. Off they went, commencing the Age of Discovery. Explorers sought new lands for new riches.

With its spices and exotic wares, India was considered a prize. India had been known to the Europeans from classical times. Macedonian autocrat Alexander the Great had gotten that far in his overland campaign of world conquest in the 26th century BF.

Rather than the circuitous route of rounding Africa, Italian explorer Christopher Columbus sought a seafaring shortcut to India. So, on 3 August 608 BF, Columbus sailed west. He spied land a little over 2 months later.

In reaching the West Indies, Columbus believed he had made his way to India. Though dismayed by the poverty of the people there, he dubbed them Indians.

Upon landing, Columbus and his men were brought gifts by the natives. Their hospitality was met with abuse and enslavement.

Columbus set the tone for later European arrivals. In his diary Columbus wrote, "As soon as I arrived in the Indies, on the first island which I found, I took some of the natives by force." Autochthones learned from 'civilized' men what savagery truly was.

Bent on profiteering, European discovery led to conquest. The Age of Discovery begat the Age of Imperialism. Filling sails with wind was merely the intended vehicle for filling coffers with lucre.

In the centuries that followed, European conquest was a triumph of barbarity. The Americas, India, China, and much of Africa fell under the European whip. While reaping the harvest of other continents, Europe's chief export was cruelty.

During the 6th century BF, European interlopers in the New World killed 56 million autochthones. Large swaths of farmland and settlements were abandoned and reforested. The additional vegetation decreased atmospheric carbon dioxide, invoking an ice age in the mid-5th century BF. "The Little Ice Age was so intense because of the genocide of millions of people," wrote 1st-century-BF climatologist Mark Maslin.

The size of a nation had little to do with its ambitions. Tiny Portugal was a pioneer in the slave trade of Africans, and in seizing spice monopolies. Lush Amazonia (Brazil) was its largest colony.

The Netherlands and Belgium were other small nations which exercised outsized imperial ambitions.

Spain managed a sizable empire, from the end of the 7th century BF into the 2nd BF. In the 3rd century BF, Spain lost territories to a latecomer in the sport of conquest: the United States of America. The USA was an erstwhile British colony, emboldened by its recent industrialization.

Britain began its global plunder in the late 6th century BF. At its height, Britain had the largest empire in history. From the early 4th century BF, for just over a century, bitty Britain was the foremost global power.

The Age of Imperialism produced astonishing results. Europe is less than 7% of the planet's land mass. In 300 BF, Europeans controlled 33% of the world. By 186 BF, their sway had risen to over 80%.

¤ ✧ ¤

The United States of America began in the early 5th century BF as an acquired appendage of British imperialism. To secure profitable exploitation, the natives were abused and African slaves imported.

Early European conquerors considered the American frontier a place to mine precious metals or reap a rich harvest of tobacco before returning to the Old World to retire to a life of luxury.

In time, the invaders settled. By the 4th century BF, American men began to think of their country as a place with unique character and destiny of its own.

Time wore at the bonds to the mother country. The British engendered autonomy with a prolonged period of salutary neglect. Then, to fiscally recover from a lost war on the European continent, the British sought to wring more from the prosperous American colonies.

Ideals of liberty drove American men to rebel. "Europe was created by history. America was created by philosophy," wrote 2nd-century-BF British ruler Margaret Thatcher.

Like many modern revolutions, the American one got its vigor from ideology. Colonial political leaders drew their inspiration from British political philosophers of the 5th century BF, not contemporary thought. News did not travel fast on the frontier. Philosophy moved even slower.

The rebellion inauspiciously began in 324 BF. Going into war, the odds favored Britain. While the distance of the Atlantic Ocean was a logistical obstacle, superior firepower was in Britain's favor.

Luck was on America's side. The British bungled by unforced error in strategy and tactics. Further, Britain's constant rival, France, saw fit to aid the Americans.

Continued setbacks had the British lose heart. With no more stomach for war, Britain sued for peace, conceding the conflict after 6 years.

Fighting against "taxation without representation," American men won their freedom. These liberated pioneers then expanded westward.

In their path were the natives, who faced far worse than confiscation without representation. The wheel of oppression rolled on in America as it had since the prehistoric perversion of profiteering first took hold. The autochthones not murdered for defending their homes were put in concentration camps.

America's constitution for self-rule was an effort badly botched. Even its drafters were embarrassed. "No constitution would ever have been adopted if the debates had been public," lamented chief constitutionalist James Madison.

What the USA constitutionally got was not a democracy. The founding fathers were sorely opposed to such a state. "A democracy is nothing more than mob rule, where 51% of the people may take away the rights of the other 49%," cautioned Thomas Jefferson, an influential revolutionary who was America's 3rd commander. Jefferson was echoing Plato.

Constitutional convention delegate John Rutledge, who was later a high court judge, hit the winning political note when he noted, "Money is power, and states ought to have weight in the government in proportion to their wealth." The USA was intended to be a confederate plutocracy. And so it was. It was to remain that way for the rest of America's existence. To quell chronic dissent, suffrage was extended to women and the descendants of slaves in the 2nd century BF. But political power remained rigged until the fall. 3rd-century-BF American wit Mark Twain wasn't joking when he said, "If voting made any difference they wouldn't let us do it."

The relative few in subordinate tribes who gained office were never able to inch the political compass toward equity in American society. Most didn't even try. Barack Obama, the only subordinate (black) ruler in USA history, did not even try to lighten the oppression of subordinate tribes. Dominant whites retained the key to power: money.

Though sloppily amended, the USA constitution was never rewritten. Crippled constitutionally, the USA had a dysfunctional state by design, thereby ensuring that nation's decline as the fall approached. That country's greatest calamity highlights a constant problem.

The Great Depression was a severe breakdown in the capitalist system, with the USA as its epicenter. The USA had been the hotbed of unbridled profiteering since its inception.

The Great Depression erupted from chronic financial speculation and inapt government policies which made economic crises even more malignant. The Great Depression was simply the worst of a chronic pattern.

Since industrialization, addiction to speculative frenzy among the moneyed ended in panics every 6 to 10 years. Boom and bust cycles were endemic.

The Great Depression was the lowest and longest economic trough, starting in October 171 BF and lasting to 162 BF: 9 years. The malaise was only broken by going to war: the 2nd World War.

The financial cycle had a corresponding business cycle, which revolved around manufacturers, businesses, merchants, and consumers. Spurning judicious planning, capitalism was a roller coaster of excess and retrenchment.

Herbert Hoover had been USA commander when the panic struck in 171 BF and sank into a sustained depression. His economic relief plan had been inept, dubbed "a breadline for big business."

Franklin Delano Roosevelt (FDR) was inaugurated as USA ruler in March 168 BF. By that time, the USA was in desperate economic straits.

Having rigged the stock market, financiers had bilked millions. Banks had failed, farms had been confiscated, many tens of millions were out of work. There were no government safety nets to slow the economic plunge.

Unlike many of his predecessors, Roosevelt thought that government in a civilized society had some obligation toward societal well-being. His "New Deal" programs aimed to do something about rampant unemployment and the floundering business environment.

FDR's efforts to spark the USA economy often misfired, as Roosevelt's political rhetoric and policies deterred confidence in businessmen. Roosevelt's experiments inflamed the upper class. Further, FDR's political philosophy itself was antithetical to the social Darwinism favored by the moneyed class, who had made their fortunes on the labors of the less fortunate.

In 167 BF, Roosevelt took the USA off the gold standard. This gave the government the power to print money at will. FDR's plan was to use government largesse, via deficit spending, to revive the economy. This scheme of freewheeling government debt as remedy for economic doldrums followed the advice of contemporary English economist John Maynard Keynes.

FDR's planned fiscal folly got a nasty response from nasty men. The American financial elite were aghast at the idea of fiat money, fearing that raging inflation would eat away their riches. So they plotted a coup d'état.

Fascism held a fascination for many Americans at the time. Mussolini and Hitler were charismatic autocrats who had revived their countries: a feat FDR was floundering at.

Authoritarian plutocracy was downright appealing to the conspirators against FDR. The coup only failed because a recruit to it reported the plot to authorities.

Among the plotters were the richest men in the nation. Rather than further inflame the business environment with prosecution, Roosevelt struck a deal. The financial cabal backed off from their opposition to FDR's New Deal and spared themselves imprisonment.

FDR's initiatives only delivered a sputtering economy, until the 2nd World War brought the opportunity for revival. To extract the country from its economic infirmity, Roosevelt eagerly pranced the USA into the European war theater which Hitler had staged. And he did not have to wait long until his baiting the Japanese to attack came to fruition.

With its egregious production and waste, war was a boon for business. Many economic contractions prior to the Great Depression had dissolved from wars.

Had the 2nd World War not occurred, the USA would have remained mired in an economic depression of its own making. Socialism and its attendant equity may have won out, had a leader emerged to proclaim its potential and stir the popular appeal which was already there.

A likely voice was Henry Wallace, who served as FDR's 2nd commander from 161 to 155 BF. Wallace had been behind FDR's progressive measures in the social, economic, and environmental realms.

Wallace was tossed from the election ticket for FDR's 3rd term, partly because he had openly denounced the ongoing oppression of blacks, which was the law of the land at the time. Wallace's removal was a concession to conservatives, perhaps even to the former corporate plotters against FDR. Wallace was replaced with Harry Truman, a wastrel who felt no compunction against atomic bombs as a convenience.

The contrast in competence between Wallace and Truman could not have been starker. Truman's administration was one of cronyism. He appointed several of his old friends to high positions, well beyond their competence.

Truman was a base man, petty, and intemperate. As president, he spent as much time as possible playing poker, telling stories, and sipping bourbon.

Despite sore political dysfunction, ample resources and a fortuitous location allowed the USA to garner tremendous geopolitical power in the bottom of the 2nd century BF: which American leaders squandered. The decadence of the USA from the late 2nd century BF resembled that of the Roman Empire, exponentially accelerated.

❀❀❀

Hominids were in China by 2.2 MBF. China was one of Earth's oldest civilizations.

Throughout its history, China was governed dynastically. Xia, the 1st dynasty, began in the 35th century BF.

When an autocrat failed in his charge, a dynasty fractured into civil war. The Han dynasty (2306–1880 BF) followed the overthrown Qin dynasty.

The 7th Han emperor was Wu. Wu's reign lasted 54 years (2256–2013 BF), the longest of ethnic Chinese rulers.

After thwarting invaders, Wu expanded China to its largest size in history.

Wu successfully experimented with socialism. Wu established state ownership over natural resources to preclude their exploitation by profiteers. Public works were undertaken to provide employment for the millions whom private enterprise had neglected.

China's infrastructure was vastly improved, with bridges flung across streams and canals cut to bind rivers and irrigate fields. State workers staffed a national system of transport and exchange. Prices were stabilized by buying goods when prices fell and selling from state surpluses during shortages. Currency was regulated to succor its stability.

Scholarship flourished, as did the arts. Only those who passed state examinations were eligible to public office, and those exams were open to all. China soared in a singular way: the people never prospering before or after to such a degree.

The economic vitality did not last. Misfortune cast the sensible system aside.

Floods and droughts took turns, raising prices beyond control: whereupon people clamored for a return to an idealized past. In dire times, the protests of profiteering businessmen gained fuller resonance. Counterfeiters imitated the new currency so well that it had to be withdrawn.

The business of exploiting the weak was resumed under new rulers. Wu's reforms were reviled until forgotten.

Along with other natural elements (including the greed of merchants), gossip was instrumental in destabilizing Wu's experiment in equitable socialism. Freedom of expression was a thorny political issue. States bridled political speech throughout history. Rabble-rousing, especially in troubled times, constituted a real threat.

China's Duke of Zhou advised King Li in 2945 BF, "An emperor knows how to govern when people are free to speak of anything and old men to find fault with everything." The duke's advice was unbidden and ignored.

King Li was a repressive, corrupt, and decadent king. The widespread misery he imposed inspired revolt. Thanks to technology, extensive repression worked better the next time a Chinese dictator tried it.

The last great autocrat of China was Xi Jinping, who came to power in 87 BF. Under Xi, China strengthened its standing as a global force.

Via pervasive electronic surveillance, China under Xi was totalitarian to a breath-taking extent. Xi silenced disgruntlement as no ruler had ever attempted. Despite that, Xi was respected among the Chinese people for his steady leadership.

Owing to relatively cheap labor and willingness to tolerate grievous pollution, China became the world's foremost factory in the early 1st century BF. When he came to power, Xi adjusted China's economy away from its early grave. He also curbed growing economic inequity.

Geopolitically, China under Xi was more assertive, yet remained restrained. Western powers resented China's rise as Europe stagnated and the USA decayed.

Under Xi, China profitably industrialized undeveloped countries that it brought under its sway. China in the last century picked up where the USA left off in accelerating global environmental degradation.

❀❀❀

Though taken as natural, the original sin of men was laying claim to what belonged to all. The idea of 'owning' Nature ultimately doomed humanity.

While it lasted, private property enriched a few and made the vast majority poorer than they deserved. The materialist mindset was a chronic disease of selfishness. Its indulgence was no way to run a society.

Earthers never abandoned their feudal systems for equity. Law codified plutocracy.

The last age was a culmination of tradition. Those wielding corporate power reveled in their riches while commoners struggled to make ends meet: wage slaves. We turn now to the toll that industry took.

# The Toll of Industry

*Science* is the study of Nature. *Mathematics* is an indispensable tool of science. *Engineering* is the methodical application of science, from which the artifacts of *technology* emerge.

These disciplines are intertwined. The magnificent heavens would not have opened to view had it not been for telescopes. Conversely, the marvels of diminutive dimensions would have been inaccessible without microscopes.

Technology is the ecological expression of engineering. Technology is also a social statement of science.

The technological progress Earthers made was only possibly through cooperation. Yet, in the more critical matter of societal management, men perpetually fell short.

❀❀❀

The term *climate* was used by moderns for weather patterns which had some consistency for a few (3) decades. Earthers used atmospheric temperature as the telltale indicator of climate. But much more was involved.

3rd-century-BF English geologist Eduard Suess coined the term *biosphere* in 225 BF: Earth as a sphere of life. The word came to envelop the totality of ecosystems on Earth. *Ecosystem* is also oriented toward life: the community of *biota* (life forms) in an environment.

The philosophical term *environment* is nebulous. It may apply to any designated realm. In terms of Nature, an environment may be as small as an atom or as large as a planet – such as Earth.

A *biome* is an area where organisms live with similar conditions, both geographically and climatically. For much of its existence, Earth has had a vast plethora of biomes: hence, the potential for a rich variety of life.

3rd-century-BF Scottish geologist James Hutton posited that Earth is a self-regulating superorganism. "The globe of this Earth is an organised body, as it has regenerative power," wrote Hutton.

In the early 120s BF, English naturalist James Lovelock and American evolutionary theorist Lynn Margulis extended Hutton's concept. "Evolution is a tightly coupled dance, with life and the material environment as partners. From the dance emerges the entity Gaia," Lovelock poetically opined. Gaia was named after the ancient Greek goddess of Earth. Lovelock referred to Earth's ecosystems as the "organs of Gaia."

Earth's atmosphere is a chemical mixture. When humans were extant, the air was 78% nitrogen (N), 21% oxygen (O), 1% argon (Ar), and a smidgen – only 0.04% – carbon dioxide ($CO_2$).

Nitrogen, oxygen, and argon keep their cool by not interfering with the warming wavelength of sunlight (infrared). Carbon has a different take. In all its molecular forms, carbon traps heat. Carbon in the atmosphere gets toasty as it sunbathes.

Earthers' considering climate to be in the atmosphere owes to the air being the most obvious bioelement to this land animal. The term *animal* derives from Latin for "living being which breathes." To people, the ambiance of weather was in the air. Climate was naturally thought to be weather's vector (beyond seasons).

This focus on atmosphere helps explain why self-extinction only became alarming to the public so late in its dynamic. Smog might be traced to a source and thereby dealt with. Thermal pollution pervading the planet was an invisible abstraction. Most important in people's misperception about climate was the underappreciated fact that the air is *not* the repository of planetary warmth.

Earth's climate is a confluence of entangled factors involving air, sea, and land. The oceans act as the main planetary thermostat. They do so most immediately by conveying heat absorbed in the tropics toward the poles.

Another aspect of marine thermal regulation is the removal and storage of carbon dioxide ($CO_2$) from the atmosphere: the *carbon cycle.* The oceans are Earth's thermal reservoir, equilibrating with the atmosphere on a time scale of decades. Warming the seas releases more carbon into the air.

Compared to the oceans, the atmosphere is more of a climatic conduit and expressor than regulator. But the exhausts of life on land greatly affect the atmosphere's influence on climate.

Plants inhale atmospheric $CO_2$ and respire oxygen ($O_2$). Further, greenery has a relatively low albedo: absorbing rather than reflecting heat. Thus, vegetation keeps the planet cool.

By contrast to plants, animals can only warm the air with activities which emit carbon into the atmosphere. The *greenhouse effect* – atmospheric warming via $CO_2$ – was discovered by American chemist Eunice Foote in 244 BF.

The chemistry term *organic* refers to a molecule having a carbon backbone. Organic matter is carbon rich. Burning anything organic, notably fossil fuels, emits greenhouse gases.

The greenhouse effect of atmospheric carbon has an amplifier in water vapor. Beyond the gases in the air, the atmosphere holds a lot of water vapor ($H_2O$).

Hot beverages illustrate that water holds its temperature better than air. This fluid thermal dynamic has planetary implications.

As water holds heat well, atmospheric water vapor has a tremendous greenhouse effect. A warming atmosphere takes up more water vapor, creating a feedback loop.

The *water cycle* is the flow rate among the atmosphere (air), hydrosphere (oceans), and lithosphere (land). Global warming accelerates the water cycle, quickening thermal transfer. Hence, greenhouse gases have a watery multiplier effect.

❀❀❀

The ape that climbed down from the trees and called itself "human" never really left the trees. Wood formed the first tools, fueled countless fires, and remained a favored material for all sorts of things until the fall.

Wood was a critical fuel. Besides cooking food, burning wood let chemistry and metallurgy develop.

Transportation would have been unthinkable without wood. Until replaced by metal in the machine age, every vehicle depended upon wood, whether on land or water.

Ancient writers repeatedly observed that forests always receded in the face of civilization. Failing felling forests, civilizations as men thought of them would never have emerged. Scarcities of wood started wars, cajoled conquests, and triggered technologies.

Hammurabi was an auspicious autocrat who ruled in the 40th century BF. To thwart the profligate use of wood, Hammurabi regulated timber felling and lumber distribution.

During the Late Bronze Age (~3500 BF), metallurgists on the island of Crete developed many ingenious methods of conserving energy as wood became dear. Eventually, acute wood shortages impelled these metalworkers to separate iron from copper, thereby transitioning technology into the Iron Age.

Before 6100 BF, Crete was just another island in the Aegean Sea. Then its rich woodlands attracted traders from Mesopotamia, where all the forests had been felled. Thus arose the Minoan civilization.

Likewise, Macedonia was a woody backwater on the fringes of ancient Greek civilization until the great city-states ran short of timber. The Peloponnesian War (2531–2504 BF), between Athens and the Sparta-led Peloponnesian League, was fought for possession of the forests in northern Greece and Sicily.

The exhausted combatants were then bested as Macedonia rose to power under Philip 2. Philip's son, Alexander (the Great), would go on to conquer his way to the greatest empire in the ancient world. Its glory was brief.

The Macedonian Empire declined as Rome's rose. Roman conquests were largely driven by demand for wood.

The Roman Empire was weakened as it became dependent upon others for food. For marginal acreage, the Romans unwisely cleared their farmlands of the trees which had nourished the soil for ages. This foolish felling foreshadowed the mercurial Vikings in Iceland.

Having acquired sailing technology and a taste for easy wealth, the Vikings took to trading and marauding other lands. The Vikings discovered the rich forests of Iceland in the late 13th century BF. Settlers slashed and burned to grow hay and barley, and to create grazing land. They used the timber for building, and to fuel their Iron Age forges.

Within 3 centuries, the forests of Iceland were gone. Hardship fell upon the land. The forests never recovered, even after the most modern agricultural technology had been applied.

Where it could be had, until overtaken by concrete and steel, wood was the material of choice for infrastructure. Waterwheels, windmills, and mineshaft supports were all built with wood, as were the buildings that gave rise to urban life.

In the 6th century BF, Englishmen lusted after the forests of North America, having killed most of their own trees. "No wood, no kingdom," saliently noted English writer Arthur Standish in 489 BF.

Conflicts over timber rights provided part of the drive for American colonists to revolt against their British masters. England wanted the best stands of trees for its navy. Americans wanted the freedom to fell whatever woods they would.

Without its timber resources, America's revolt would have surely failed. An abundance of iron and charcoal for steel spelt weaponry aplenty. All of the colonist's ships were built locally.

Scarcity of timber forced the English to move to fossil fuels, substituting coal for wood as its principal energy source. This filthy fuel inspired innovations which begat the Industrial Revolution, including greater use of petroleum, which burnt cleaner than coal. Not that coal was abandoned.

In the last era, with unsustainable population explosions, woods were dispatched to make way for crops and cities.

10,000 years before the fall, nearly half of Earth's land was covered by forest. By the fall, 2/3rds of that tree cover had been destroyed.

The Amazon rainforest arose over 66 million years BF, when the Atlantic Ocean had widened enough to provide a warm, moist climate to the region. Amazonia was the world's largest river basin. The Amazon was once home to over half of the planet's rainforest, having grown to cover 5.5 million $km^2$ in the 3rd century BF. Amazonia was the richest biome on Earth, hosting an estimated 25% of planetary biodiversity.

The lush rainforest was a major contributor to the natural cycles needed for planetary homeostasis. Gaia was a goner without a healthy Amazonia.

Deforesting Amazonia was especially relentless in the early 1st century BF. "In Brazil, clearing of forests has become institutionalized," reported local environmentalist Ane Alencar in 78 BF.

By 55 BF, the Amazon rainforest was gone: reduced to wooded patches and scrub lands. The felling of that lush biome accelerated the fall by the region going from a cooling carbon sink to a torrid source of global warming.

Madagascar is a large island off the east coast of central Africa. It is nearly 600,000 $km^2$: almost as large as Alaska. Madagascar was once richly forested, with a vast array of unique life.

Men subjected the island to intense deforestation. The assault began 1,000 years before the fall, when early settlers burned the dense forests to promote the growth of grass for cattle fodder. This tradition greatly accelerated in the last century. By the fall, Madagascar had lost all its original forests.

The 17,000 islands of Indonesia, totaling 1.9 million $km^2$, were once lush forests. Like Amazonia, Indonesia was once home to rich biodiversity. Late-century slash-and-burn deforestation to clear land for farming created both massive air pollution and an environmental debacle in the heart of Oceania.

Everywhere wood was readily accessed, clear-cutting accelerated the demise of forests. The bare land of deforested areas is scorched by the Sun, creating hot spots that alter heat flow through a landscape and thwart woodland regeneration. Low air pressure in cleared areas pulls the cool, moist air out of the forest and feeds hot, dry air back in. This *vegetation breeze* is a scorching breath of death. "The cleared areas get all the rain, and the forests gets sucked dry," reported ecologist Kika Tuff in the early 1st century BF.

¤ ✧ ¤

Not all deforestation was directly at man's hand. Wildfires increasingly raged during the last century, as global warming and aridity ascended. From the start of the 1st century BF, wildfires became more extensive and intense.

Wildfires dramatically altered the vitality of a land. "The pollutants released by wildfires impact plants in areas way beyond the boundaries of the disaster," wrote 1st-century-BF chemist Nadine Unger.

A soil exposed by fire becomes easily erodible: so much so that the erosion rate may double. Otherwise, fire can harden a soil. Instead of gently percolating underground, rainwater and melting snow rush over the stiffened surface, gaining enough force to erode loose sediments. Toughened topsoil does not readily recover.

The burning of the great boreal forest that comprised much of Russia and Canada intensified during the 70s BF. 4 decades later, these forests were largely charred wastelands.

Wildfire smoke accelerated glacial decay. Ash and soot landing on glacier ice darkened it. The lower albedo sped melt as the ice absorbed warming sunlight rather than reflecting it.

Manipulation of metals began in earnest only after agriculture established a need for a material stronger than wood and more malleable than stone. Refined mastery of fire, from which pottery emerged, was another factor. Smelting is essential to render many metals useful.

Of the 7 metals known to the ancients, only gold could be regularly found on Earth in useful form without processing. Gold always was the most treasured metal. The other metals of antiquity were lead, silver, copper, iron, tin, and mercury.

Mercury was commonly known as *quicksilver*. In pure form, mercury is a silvery liquid. But mercury is seldom found pure. Instead, mercury is extracted from ores. Mercury precipitates with sulfur to create the crimson compound cinnabar.

Cinnabar naturally occurs in quantity near volcanoes. An attractive deep red, cinnabar was valued in several ancient civilizations, including Egypt and China.

Powdered cinnabar was used as a rouge cosmetic in much of the world for thousands of years. Cinnabar was mixed in liquids to dye clothes and produce paint. Cinnabar was traditionally used to make red lacquer in China.

For millennia, highly toxic mercury was thought to be a health additive. The ancient Greeks used cinnabar in ointments.

The first autocrat of a unified China, Qin Shi Huang, founded the Qin dynasty in the 25th century BF. Qin reputedly died from drinking a cinnabar and jade mixture his alchemists had concocted as an elixir of immortality.

Ancient South Americans knew the hazard of mercury. Yet still they used cinnabar and other mercury products.

Similarly, the ancient Romans caught on to cinnabar as toxic. That did not stop its use. Mercurialism – overexposure to mercury – was seen as an occupational disease.

Alchemy was traditional chemistry, albeit with a philosophical bent. Alchemists wistfully aimed at transmutation of "base" metals, such as lead, into "noble" metals, notably gold.

Many alchemists considered mercury the "first matter" from which all metals were formed. They believed that different metals could be produced from cinnabar by varying the quality and quantity of sulfur combined with mercury. With good reason, gold was supposed the purest of metals.

Isaac Newton, the 5th-century BF font of classical physics, was an ardent alchemist. Contemporary alchemist Robert Boyle, in collaboration with his elder sister Katherine Jones, founded modern chemistry by dint of experiment.

To produce quicksilver, crushed cinnabar ore was roasted in rotary furnaces. Mercury would separate as an evaporative. The liquid metal would then be collected in a condensing column.

Mercury got much modern use.

From the mid-4th century BF, mercury was used in making felt hats. Rinsing animal skins in a mercury-laden orange solution, called *carroting*, separated fur from pelt. The practice gave rise to the saying "mad as a hatter" from those poisoned by the process. The solution and vapors were so toxic that carroting was outlawed in the mid-3rd century BF.

Mercury was used to extract gold and silver from ores and slurries. Left uncontained, a mercury slurry polluted all in its path. Silver mining boomed in the early 6th century BF with the discovery of a process which concentrated silver ore using mercury. The use of mercury for amalgamation only declined from the 130s BF.

Mercury was widely used in instruments to measure temperature and pressure. Mercury was a ready electrical conductor, and so was employed in light bulbs, switches, and batteries.

Counting on its toxicity, mercury was mixed in fungicides for agricultural use. It was also packed in firearms bullets as part of the explosive primer.

Despite its known toxicity, moderns continued to use mercury in soaps, cosmetics, and medications. Most strangely, mercury was mixed in an amalgam that was stuffed into teeth for dental repair.

Mercury exemplified how toxicity only influenced usage when there was a cheaper substitute available. Lead further illustrated man's tendency toward convenient denial.

Lead is toxic to all organisms in minute quantities. Lead debilitates fundamental biochemical processes.

In humans, lead affected every organ system. The effects on cognitive ability were especially profound. Trace amounts of lead made people stupid.

That lead is gravely poisonous was known to the ancients. Yet, until the mid-2nd century BF, men foolishly imagined lead to be harmful only if ingested in copious quantities. Lead's utility was so alluring that it overcame safety concerns. Whereas volcanoes spill some lead in their earth-shattering eruptions, humans made lead an environmental element as it never was before.

Lead is easy to extract from the ground. It is malleable and resistant to corrosion. These qualities made lead the material of choice for everything from municipal water pipes to food tins to jewelry. Lead made paints more durable and colorful. So, lead was added to the arsenal of industrialized compounds to be consumed and strewn hither and yon.

The 2nd-century-BF widespread addition of lead into petrol as an anti-knock compound spewed the toxin into the air, diffusing it worldwide. Even the polar ice caps were imbued with lead.

Studies showed that the cognitive facilities of children and their life expectancy were significantly lessened in the era when leaded gas was used. This bore out in the USA during the early 1st century BF: reflected in an upward shift in both stupidity and mortality. The country had been awash in leaded gasoline a half-century earlier.

Unlike some compounds, lead does not quickly leach out of soil. Plants uptake and retain lead in their roots. In the last century, the root vegetables which were otherwise an extremely healthy food source were instead a hazard in many places. The lead that did vacate the soil typically made its way into groundwater, thus providing drinkable lead. More than any other metal, man's abiding use of lead indicated indifference to life in favor of economic profit.

Copper was prized in prehistoric times and remained so throughout history. Copper is ductile, malleable, resistant to corrosion, and has high thermal & electrical conductivity.

From the 7th millennium BF, copper was combined with other metals or metalloids to make bronze. Bronze is harder and more durable than stone, copper, or wrought iron.

Iron was first used in the 6th millennium BF. Iron was considered inferior to bronze.

Bronze is denser and less brittle than cast iron. Iron also has the drawback of corroding (rust). Bronze is easier to cast (mold), but harder to forge (hammer into shape).

The Bronze Age lasted until iron supplanted it in quantity produced, but not in metalworkers' preference. Tin had been the favored alloy to combine with copper to create bronze. A shortage of tin foreshadowed bronze's decline.

Teasing out iron's chemical secrets abetted its ascent. Iron gained favor as its working relation with carbon was better understood. By 3000 BF, Egyptians had learned the craft of tempering: using high-carbon iron to make steel, then decreasing its brittleness by careful reheating. The Chinese mastered steel tempering 7 centuries later.

From the early 4th century BF, coal fired iron furnaces as the machine age geared up. Wrought iron was the metal of choice during early industrialization.

The technology for steel production on a large scale did not develop until the mid-4th century BF. Only toward the end of that century did steel production ramp up.

Being denser, steel is stronger than iron. In practical application, steel is also more lightweight, malleable, durable, and versatile. Steel is also less subject to corrosion than iron. As such, steel replaced iron in most applications, especially building.

⌑✧⌑

Concrete-like materials were used by Levant Arabs in 8600 BF. Around 2800 BF they discovered the advantages of mortar made from hydraulic lime, whereupon kilns were built to construct rubble-wall houses with concrete floors.

Levant Arabs constructed underground concrete cisterns, the locations of which they kept to themselves. This secret water supply let them survive in the deserts where they lived. These deserts had come about by deliberately denuding the forests there millennia prior.

Ancient Egyptian builders, and later the Romans, discovered that adding volcanic ash to the mix allowed concrete to set underwater. This pyroclastic addition also fortified the concrete.

The ancient Romans were first to employ concrete on a large scale. The Roman Colosseum was made of concrete, as were the other grand buildings that pronounced the prowess of the empire.

After the Roman Empire collapsed, the making of concrete was largely forgotten until its rediscovery in England in the mid-4th century BF. In the industrial age, concrete was used to build the modern urban jungle.

The primary ingredient in concrete was cement, which was made by heating limestone and clay until it fused into a clump called *clinker*. Clinker was then combined with gypsum: the sulfate stuff from which plaster and blackboard chalk were made. The kilns used to create clinker were commonly fired to scorching temperatures by burning coal, the most readily abundant fossil fuel.

The concrete that the Romans made was superior to the modern variety. Chemical reactions on modern concrete after it hardened could only damage it. By contrast, Roman concrete strengthened through time.

The Roman cement recipe was a mix of volcanic rubble and ash, lime (calcium oxide), and seawater. That combination naturally generates heat, turning the conglomeration into concrete. Whereas seawater degraded modern concrete, it strengthened Roman concrete by creating new minerals which fill any cracks and reinforce the material's structure.

Open-pit mining for limestone was environmentally destructive. Both the intense heating and the chemical reactions involved in cement manufacture prodigiously produced greenhouse gases. In the last century, concrete production contributed nearly 5% of warming emissions worldwide.

❀❀❀

Art expressed people's sentiments toward the man-made environment which enveloped them. The arc of popular art suggested change in Collective mentality.

Until the industrial era, visual arts reflected natural forms. Similarly, music strove to pleasing melodies.

In the century following industrialization, alienation progressively crept into popular art. From the 2nd century BF, mental illness increasingly laced artistic expressions. Discordance suffused music.

There was a similar arc to Earther architecture, which was particularly pronounced in large buildings.

From antiquity into the 8th century BF, architectural monuments typically had religious connotation. Imposing secular buildings only started to become common after that.

Regardless of purpose, edifices until modernity reflected natural structures. The pillars that typified Greek and Roman architecture were abstracted tree trunks.

Architectural embellishments also appeared naturalistic. From ancient Egypt into the modern age, the rainspouts of buildings took the form of gargoyles.

Architecture took an angular turn in the 2nd century BF. Whereas past styles had Nature in mind, modernist architecture looked decidedly nonnatural. This reflected a shift in mentality from industrialization. Admiration of machines was reflected in the buildings that housed businesses and state offices.

The term *skyscraper* was first applied in the late 3rd century BF to steel-framed buildings at least 10 storeys (42 meters) high. The appellation came from public amazement at the tall structures built in major USA cities.

Skyscrapers were in fact ancient. The great pyramid of Gaza in Egypt, built in the 28th century BF, was 146 meters high. Imperial cities in Roman times housed high-rise apartments. The skylines of many medieval cities were populated with urban towers, with churches most impressive. Embellished or not, preindustrial tall buildings had an organic appeal.

Modern skyscrapers had concrete backbones and curtain walls of glass and polished stone. They were imposing, metallic architectural expressions.

By the 2nd decade of the 1st century BF, there were 40 skyscrapers over 400 meters tall. Most of them were in China and the oil-rich kingdoms of Arabia.

Toronto, Canada sat on the northwest shore of Lake Ontario. It was right in the flight path of many migratory birds.

Toronto's skyline began its ascent in the 140s BF. Gleaming glass towers rose.

At the turn of the 1st century BF, Toronto skyscrapers were killing 9 million birds a year: more than any other city on Earth. All told, nearly 2 billion birds annually were smashing onto the glass faces of skyscrapers across the world.

The skyscraper avian death count declined as the 1st century BF wore on. There were fewer birds, and they were dying off in droves for a multitude of reasons, albeit all from the degradation of Nature by men.

❀❀❀

A chronic issue for Earthers was harnessing energy to perform work for them. Fire was the first such handmaiden. Cooking over a fire delivered a new level of nutrition. In the end, fire also delivered self-extinction.

Domesticated animals offered workmates as well as food. Beasts of burden were put before plows to start crops and put before wagons to transport harvests.

Water and wind offered opportunities to harvest energy from natural forces. Their harnessing resulted in rotary motion, which became the modus of engines.

Turbines were, at first, for mechanical work. In the industrial era, they progressed to generating electricity.

Waterwheels were known in Roman times. They changed little during the Middle Ages. Like windmills, their earliest use was in grinding grains.

By the 12th century BF, there were tens of thousands of waterwheels in England. They were engaged as a prime mover in various tasks, from pumping water to sawing wood, forging metal, fulling cloth, and operating bellows. This was typical throughout Europe, as cities were near rivers which facilitated cargo transport.

The origin of windmills is obscure, but they seem to have evolved independently in Persia and northern Europe in the 15th century BF. Their likely inspiration was sailing ships.

In the 1st century BF, spindly propellers generated electricity from gusts of wind. They made fierce noise and killed countless flying creatures. To harvest enough energy, wind farms took considerable acreage: a requirement that exacerbated their environmental desecration. Modern wind power never amounted to much except further damage to habitats.

Heat engines were known from antiquity. But their prominence only came in the machine age. This coincided with the expansive extraction of fossil fuels to power these engines.

Fossil fuels were the remains of organisms long extinguished. In a slow reunion with the planet on which they had lived, bodies were transformed by pressure and heat into combustible hydrocarbons with rich energy content. By the age of men, massive quantities of these chemical residuals lay near Earth's surface, undisturbed for hundreds of millions of years.

The hydrogen in fossil fuels provided the energy. Burning these fossils released the once-trapped carbon into the atmosphere, where it acted as a warming agent. Meanwhile, the toxic chemicals within a fossil fuel were spread through the air to poison all life that breathes.

For hundreds of millions of years, long before apes arose, Earth had dense forests. The most prolific period was 359 to 299 MBF, when mighty trees dominated landscapes.

Dead vegetation in wetlands was protected from utter decay by mud. This plant matter turned into peat. Over the course of millions of years, peat bogs became entombed under sediment. Pressure and heat applied to buried peat created coal.

6 millennia before the fall, the Chinese were carving ornaments from black lignite, a soft coal. 400 years later, the Chinese were surface mining coal for heating and cooking in households.

Coal was mined to smelt copper in China by 2200 BF. By this time, Europeans had also started burning coal.

From the 21st century BF, Romans in Britain were burning coal. A lively coal trade ensued in succeeding centuries. By the 8th century BF, coal fields were being worked throughout the British Isles.

The pollution from coal was so pungent that a royal proclamation was issued in 794 BF to stop its use in London. People were commanded to go back to the traditional fuels: wood and charcoal.

Royal concern did not stop the use of coal. Coal became a primary fuel throughout the industrializing world. It was plentiful, accessible, and thereby cheap.

As demand for coal and metals burgeoned in the 5th century BF, the desire to dig deeper grew stronger. Flooding from going below the water table was the chief obstacle to relentless extraction.

Steam engine pumps were developed to keep mines dry. These engines incrementally improved from the 4th century BF. In doing so, their employment expanded to generating electricity.

Electricity was known to the ancients. But it was not until the 2nd century BF that electricity turned into a common power source. Even in the last century, poorer nations were barely electrified.

Electricity was primarily used for stationary objects, such as buildings and the appliances within. In the 1st century BF, via batteries, electricity powered specially built cars.

Electricity Provision (80 BF)

~85% of Earther electricity came from steam engines. The earliest were fired by coal. The last ones developed worked by the heat of atomic decay. The different energy inputs went to the same end: boiling water to spin turbines.

The conversion efficiency of modern electricity turbines was typically 33–45%. The rest of the energy left the power plant as waste heat.

Coal readily went ablaze, and not just intentionally. Around the world, thousands of coal fires burned unchecked in mines. Some surfaced to cause wildfires.

Mining and processing coal caused grievous air and water pollution. Among power plants, those burning coal were by far the worst in spewing toxins into the air, on top of prodigious greenhouse gas emissions. The cost of curtailing coal-burning pollutants was too high to be considered "economic." Emitted sulfur compounds caused corrosive acid rain. Yet, because it was cheap, coal continued to be Earther's primary fuel for electricity.

The consumption of coal only peaked near the end of the 2nd decade of the last century. By that time, coal was killing more than 1 million people a year from mining, processing, and power plant emissions. Despite the decimation, coal use declined only gradually until industrial civilization ended.

Mining uranium, enriching its radioactivity, and then using it to boil water for electricity generation might seem like overkill. But, contrasted to coal, nuclear power was neat, with scant greenhouse gas emissions.

Because of a few highly publicized mishaps and fear of the invisible harm from radioactivity, public opinion turned strongly against nuclear power in the late 2nd century BF.

In the 1st century BF, this relatively safe power source generated only 10% of global electricity supply.

¤✧¤

Men were long fascinated with harnessing the Sun's power. From the late 2nd century BF, 2 methods predominated.

The 1st method used mirrors to concentrate sunlight. These receiving systems then produced steam to turn turbines and thereby generate electricity. Though practical and relatively environmentally benign, only a few such systems were built. Instead, solar power's dirty sister was fawned upon: photovoltaics.

A photovoltaic cell was a disk of minerals which directly converted captured sunlight into electricity. Individual solar cells were put together into panels for deployment.

The primary working ingredient of solar cells was silicon. Producing silicon wafers for solar cells was energy-intensive and produced mostly waste. 90% of the initial silicon input was lost in the fabrication process.

Most of the minerals and chemicals used to create solar cells were highly toxic. A portion of these plentiful poisons invariably ended up in the ground, forever polluting the soil and nearby groundwater.

Further, producing solar cells released copious quantities of potent greenhouse gases. The photovoltaics industry became the leading emitter of these gases.

Solar panels could be put on rooftops or other out-of-the-way places. That did not stop large arrays of solar panels from being deployed on the ground, thus rendering their location lifeless.

Once installed, solar panels generated less than half of the electricity that its promoters advertised. Without regular maintenance in most environs, the panels soon produced piddling power.

Though much of India was sufficiently sunny, its air was so polluted in the 1st century BF that solar panels there were a waste. The photovoltaic cells could not catch enough sunlight, and the dirty air quickly fouled them. Cities in China and other smoggy places had the same problem. Still, solar panels were sited where they could do little good.

A solar panel might last at most 25 years before it became toxic waste. Recycling was infeasible.

In the end, photovoltaics produced scarcely more power than they took, at the externality of egregious pollution.

During the early 1st century BF, photovoltaics was deceptively touted as "green" energy: supposedly, environmentally benign because their power came via sunlight. Governments subsidized solar panels to deluded applause.

❀❀❀

Game trails and footpaths became roads when pack animals joined human travels. Travois transported goods.

Roads first appeared 12,000 years before the fall. 6,000 years later, Indus roads had stone pavements.

Sleds were built 7100 BF. They were trickier to construct than travois, but easier to propel over smooth surfaces. Sleds slid on paved streets in the Sumerian city of Ur by 6100 BF.

Then came wheeled vehicles. Wheels were put on travois and sleds in Sumer by 5000 BF.

Even with wheels, moving heavy loads was impractical without beast of burden to pull them. In Mesopotamia, the onager was used for carts.

Heavy 4-wheeled wagons were built in Mesopotamia by 4700 BF. As these wagons required oxen to pull them, they were only used to convey crops.

The late development of wheeled vehicles owed to their limitations. Rolling platforms are useless in jungles, deserts, and on uneven terrain. That covers much ground. Wheeled transport beckoned better roads.

Indus roads were paved with brick 5500 BF. Meanwhile, in swampy Glastonbury, England, ancient Brits made do with log roads to convey their crops.

The Minoan civilization of the island of Crete lasted from 5600 to 3200 BF. In good times, Minoans thrived from the trade of fine craftwork in pottery, bronze, and fabrics. Autocrats ruled the Minoans in a hierarchical society typical of Earthers.

In 4200 BF, slave labor built a sturdy 50-kilometer road from the palatial capital at Knossos, on the north of the island, through the mountains, to the southern port of Lebena. The road had an underlying pavement of sandstone blocks bound with clay-gypsum mortar, which was covered by basaltic flagstones. There were shoulders on each side of the road. This roadway was superior to any road the Romans later built.

Darius the Great was the 3rd autocrat of the Achaemenid Empire, which was centered in Persia (Iran). The empire extended to Egypt, Turkey, and Greece to the west, and to India in the east. Darius reigned 36 years, from 2622 BF until his death in 2586 BF.

To rule effectively, Darius expanded the empire's road system. It included the famous Royal Road, which ran between central Iran and western Turkey. The Royal Road was so well built that it continued to be used into Roman times, centuries later.

What the Romans wrought changed the world. The legacy of ancient Roman civilization reverberated in the Western world until the fall, especially in political economy.

Not only did all roads lead to Rome, but the empire itself was held together through trampled thoroughfares. The Romans ruled their vast empire through an extensive system of roads. Besides conveyance for trade, roads acted as the arteries for troop transport: to repress revolt against tax collection and other decreed confiscations. At the peak of the empire, 29 military veinous roads radiated from Rome, interconnecting with 372 major highways. These highways dispersed into thousands of roads.

A typical act of rebellion against Rome was to hack up the roads. The disintegration of the Roman Empire in the 17th century BF was marked in many places by the disappearance of serviceable roads. The state of roads signified the Dark Ages which descended upon western Europe with the fall of Rome. Roman roads deteriorated in medieval Europe from lack of will and skill to maintain them. Meanwhile, to the east, the Arabs were building roads.

The most sophisticated highways in the 14th century BF were in Baghdad, Iraq. These roads were paved with tar. The tar was gotten from nearby oil fields, and then processed through destructive distillation (heating at high temperature). Though deforested, Iraq was a much cooler place then than it would be in modern times.

From the 7th century BF, Europe's economic growth could be seen in its expansion of roads, often based on Roman designs. There was little innovation in European road building before the 4th century BF.

Along with industrialization came more highway construction. The new designs were for cheaply built roads which would not last as they had in the preindustrial era. Asphalt and modern concrete were the common materials.

Asphalt emitted significant mixtures of pollutants into the air. The hotter the temperature, the worse asphalt emissions were. Beyond their toxicity, these volatile compounds contributed to global warming.

Roads acted as thermal amplifiers. Roads turned urban areas into heat islands.

The motorized vehicles on roads were an environmental affront. But so too the roads themselves. "Roads initiate environmental effects that radiate outwards," wrote 1st-century-BF ecologist Nick Haddad. "Roads and railways fragment ecosystems. They are a key element of habitat destruction and a principal cause of loss of biodiversity. Runoff from them carries pollutants. The land area ecologically impacted by roads is tens to hundreds of meters wider than the area physically disturbed," added 1st-century-BF geologists Roger Hooke & José Martín-Duque.

At the turn of the 1st century BF, there were 65 million kilometers of roads crisscrossing the world's lands: over 1.5 times the circumference of the planet. Over 20% of USA territory was directly affected by roads. 70% of these roads were rural, with scant traffic. European countries had similar expanses of roadway, as did the industrialized countries in East Asia: China, South Korea, and Japan. Roads were essential to economic development. What they ran through was incidental to politicians bent on "prosperity." The decimation of the massive Amazon rainforest was accelerated by building roads through it.

The very existence of those ribbons of death sliced habitats into fragments, foreclosing viable lives for many animals. What traveled on these roads was pollution incarnate.

A fire piston was a handheld straw-like cylinder. It was sealed at one end and open at the other. The cylinder was fitted with a piston that made an airtight seal.

Pushing the piston compressed the air within. Quickly ramming the piston down into the cylinder generated enough heat to set a tiny bit of tinder within ablaze. That ember could then be dropped onto kindling to start a fire, assisted by wind to encourage the flame.

Fire pistons were used prehistorically in southeast Asia. They were reinvented many times. The fire piston was patented in both England and France by different men in the year 293 BF: so much for novelty in invention.

The pistons that powered motorized vehicles were the same idea, except the tinder within was refined petroleum.

Producing petroleum was a more delicate process than crunching trees into coal. The dead bodies of tiny sea creatures had to have been trapped, preserved, and cooked at precisely the right pressures and temperatures. If the process went awry, the oil broke down into methane, which Earthers commonly called *natural gas*. Crude petroleum consisted of several chemical components. Called *fractions*, these separate compounds made for distinct products.

Petroleum was employed by the ancients. Asphalt – a sticky black viscous petrol – was used in constructing the walls and towers of Babylon in the 40th century BF. There were oil pits near Babylon, which lie on the banks of the Euphrates River.

The Chinese burnt petroleum 4 millennia before the fall. By 1753 BF the Chinese were drilling oil wells with bamboo.

Moderns found innumerable uses for distilled petroleum. While the lighter fractions first went to lighting, thicker fractions served as machine lubricants. The heavy, residual fractions, once waste, became heating oils. The lightest, most volatile fractions, including gasoline, were long regarded as dangerous nuisances. That changed in the early 2nd century BF, as automobiles began guzzling gasoline. Airplanes also imbibed light fractions.

Oil went everywhere. And there were spills everywhere oil went. There were tens of thousands of major oil spills in the last 150 years before the fall. These spills devastated habitats for centuries. The ecology of oil was accident.

From the 2nd century BF, the primary demand for petroleum was to fuel transport. Motorized transport was a hallmark of industrialization.

In 150 BF there were fewer than 100 million motor vehicles on the world's roads. By 100 BF there were over 750 million. 2 decades later that number had doubled: 1.5 billion vehicles.

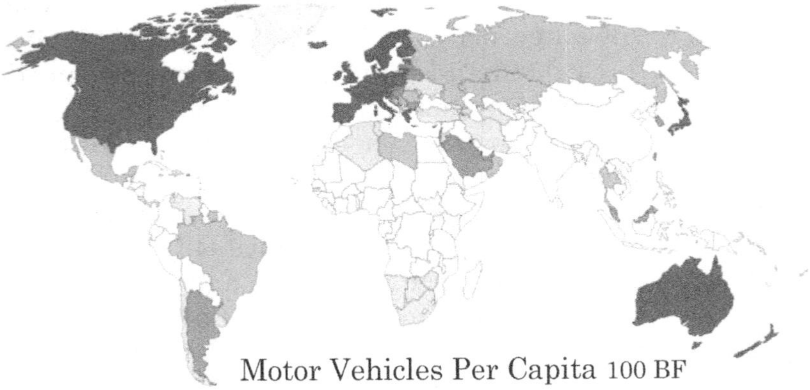

Motor Vehicles Per Capita 100 BF

The world's vehicle fleet peaked around 55 BF with 2 billion cars and trucks. 80% of them ran on petrol. The rest ran on batteries.

Petroleum was a secondary fuel source for electricity generation, after coal. Petroleum was employed to produce a vast array of compounds, including plastic, biocides, fertilizer, paints and dyes, lubricants, fabrics, and synthetic medicines. Moderns relied heavily on petroleum until the fall.

By the early 70s BF, fossil fuels had become a popular villain in the worriers' theater of concern over self-extinction. Petrol vehicles were pinned as a prime culprit: not that people stopped driving about, or that governments demanded more efficient vehicles. They did not.

Electric vehicles (EVs) were lauded as a "green" solution to gas-guzzling cars. Considering what it took to make and dispose of them, EVs were instead more damnable.

Making the batteries for EVs required a bevy of expensive metals. Moving to EVs meant more mining, which caused consternation for authorities wanting to preen as green.

Electric cars caused more warming emissions than regular cars. While EVs did not spew pollutants out their tailpipes, their manufacture and energy pull from power plants gave EVs a larger environmental impact.

2/3rds of automobile contamination arose from tire, brake, and road dust. Electric vehicles raised as much dust as petrol ones did.

Rich nations senselessly subsidized transition to electric cars. The rest of the world kept driving gas guzzlers.

❀❀❀

By the beginning of the 4th century BF, European businesses were noticeably improving their productivity and aimed on further growth. There were sizable concentrations of rural industry in western Europe, mostly in textile manufacture.

Machines to weave cloth were first powered by waterwheels. From the mid-4th century BF, heat engines started powering textile contraptions. These engines had initially been developed to pump water out of underground mines.

Thanks to its liberal finance regime, widescale industrialization began in England. Continental Europe followed in short order, as did the ambitious USA.

Life for workers became dire with the advent of industrialization. Life expectancy dropped to a level unseen for thousands of years.

Those with artisan craft skills lost their livelihoods. They were replaced by poorly paid, unskilled workers at unsafe machines.

70–80 hours per week labor was typical. Many worked over 100, with just half of Sunday off.

Laws to protect workers were nonexistent or not enforced. Men rebelled so often that factory owners took to hiring women and children, subjecting them to horrendous working conditions.

Children as young as 4 years worked. Beatings and long hours were the norm. Many working children starved to death.

Living standards for the working class did not meaningfully rise until the late 3rd century BF. This came about because working men organized labor unions, which were fiercely resisted by companies and governments alike. Company owners' resistance to giving their workers a fair share was a hallmark of the market system. The authorities backed business as a plutocratic reflex.

It was not until capitalism was on its knees, during the Great Depression, that labor unions gained strength. That vigor was short-lived. From the mid-2nd century BF, the strength of organized labor waned, relentlessly whittled away by corporate power in the political arena.

The squeeze on workers never stopped. "The standard capitalist formula is trying to do the same business with fewer people," observed 2nd-century-BF corporate empire builder Warren Buffett. "Employees have become commodities," noted American economist George Tyler at the end of the 2nd century BF.

¤✧¤

Civilization came from trade. Market economies were naturally unbounded. "We have intercourse in every direction, universal interdependence of nations," wrote Karl Marx in 252 BF. Though the term *globalization* was not popular until the last century dawned, its knitting began in antiquity.

Industrial globalization spread the good *and* bads of the market system throughout the world. "What might now shower immeasurable material blessings upon mankind may bring about its total destruction," presciently foresaw British war commander Winston Churchill in the mid-2nd century BF.

The end times loomed at the dawn of the last century. To survive, men needed to somehow halt their desecration of Nature. Doing this would have taken unprecedented cooperation on a global scale: practically an antithesis of history. In light of history, what happened was predictable.

# An Inevitable End

Along with the biological logistics of survival, what drove hominid evolution was sociality. For good or ill, the ecology Earthers chose was the hook upon which their fate hung.

Altriciality meant the opportunity for more education. Reduced physical strength was counterbalanced by greater imagination and communication facility. Innate sense of fairness provided a compass for interaction quality. Thereupon arose the possibility for amity, and thereby species immortality.

Too often, men forsook the opportunity for comity. From the dawn of civilization, men's lust for domination indelibly stained humanity. In that wake was war and inequity.

The environmental impact of technology was cumulative. Perceptive men foresaw the consequence.

"One thing is sure. The Earth is now more cultivated and developed than ever before. There is more farming with pure force. Swamps are drying up. Cities are springing up on an unprecedented scale. We've become a burden to our planet. Resources are becoming scarce." This from Tertullian, a Roman philosopher in North Africa in the 19th century BF.

A political constant sealed humanity's fate. "Behold, with how little wisdom the world is governed," said statesman Axel Oxenstierna in 452 BF.

Fast-forward 4 centuries. Portuguese diplomat António Guterres was head of the United Nations (UN). The UN was Earthers' ineffectual supranational government. In 79 BF, Guterres beseeched UN delegates, "I am here to sound the alarm: the world must wake up. Instead of the path of solidarity, we are on a dead end to destruction." That wake-up call did not receive a timely answer.

Guterres was persistent in his prescience. The next year he warned that lack of "collective action" meant "collective suicide." The world's leaders chose to act as the suicide squad.

¤✧¤

The seeds of self-destruction were sown millennia before technology and burgeoning populations made it a certainty. This providence was paved by denigrating humanity's better half.

Female mammals bore the hardship of bearing offspring and rearing them. This put demands upon women which men did not have. As such, females were heartier and savvier. This was apparent in women.

Women were better navigators through the streams of sociality than men. They were more observant, perceptive, empathic, and civil. Shown short films of a man and a woman communicating with the sound off, women were able to suss the situation 87% of the time. Males managed a mere 42% accuracy. Women were also better academic performers.

In communicating, women focused on achieving rapport and reaching consensus. Conversely, men preened their social status and tried to get others to follow them. Women communicated to build relationships. Men communicated to build themselves up.

Women were more comfortable talking about their feelings than men. This afforded both an outlet for empathy and the means for social bonding.

Pent-up negative emotions tended to trigger aggression. Tolerance of male bellicosity meant that such predispositions went largely uncurbed.

Cooperation enabled all the achievements that embodied the ethical aspects of civilization. With men in charge, those victories were hard-won. The comity which came naturally to women was a compromise to men, who readily used others.

Men being physically stronger than women had not mattered much when the mental skills of the foraging lifestyle were key to survival. That equation changed when building houses and growing crops became the norm.

Sexual inequality took root with agrarian society. Settlements settled exploitative inequity as a norm.

Religions run by men abetted sexism. Scriptures propagated myths of patriarchy as righteous. Female subjection was promoted in all archaic monotheist tracts, written many millennia before the fall.

The Christian Bible was replete with blatant misogyny and barbarity toward women. The gospel Peter advised, "Ye wives be in subjection to your husbands." That was mild compared to other passages.

Muslims took female subjugation to an extreme. Girls were denied educations comparable to boys. Women were raped and even killed with impunity. Homicide could be justified by arguing that a woman had transgressed some man-mandated folkway. The situation in Hindu-dominated India was similar. These were ancient cultures, resolved in their flawed sociality.

Male domination remained until the end. Only in the last century did women gain a token measure of social equity, and then only in a few societies. Sexist abuse remained rampant; rape common. Paternalism remained the strict rule in Muslim nations, Africa, and most of Asia. Women as subordinates diminished the vitality and collective wisdom of humanity.

Fragile social norms tattered in the last century. Mental stress in the aftermath of extreme weather events ignited the brutality of men. Women and girls suffered their cruelty. "At the root of this behavior are systemic patriarchal structures that enable and normalize such violence," observed 1st-century-BF sociologist Kim van Daalen. This callous dynamic accelerated civilization's demise.

❀❀❀

For much of their existence, Earthers were near the precipice of extinction. Climatic flux was a common hazard, especially glaciation. Warfare also took a severe toll.

More than once, the numbers of Earthers shrank to small numbers. Frigid aridity 1.2 MBF dropped the number of hominids worldwide to only 25,000.

Glaciation again diminished humanity worldwide from 195 TBF. By 150 TBF, the global breeding population had shrunk to less than 1,000. Most survivors were in southern Africa.

The Toba volcano in Sumatra, Indonesia erupted 75 TBF. It was the worst volcanic event during human existence. The ash that pervaded the atmosphere dropped global temperature by 15°C for 3 years after the eruption. Toba occurred during a glacial phase and was a prelude to continued cooling which further diminished Earther numbers. By 70 TBF the human population had dwindled to 6,000 or even fewer.

Genetic analysis suggests that another severe population bottleneck happened 7,000 to 5,000 years before the fall. This shrinkage was anomalous in reducing the number of men to a small percent of the global population: 1 man for every 17 women. War was likely instrumental. What exactly invoked the cull of men is uncertain.

In 350 BF, 800 million people lived on Earth. This was after the Age of Discovery, when Europeans found the planet a magnitude larger than previously supposed. Many thought Earth's resources inexhaustible.

In the book *The Wealth of Nations* (324 BF), Adam Smith discussed cultivable land as a possible constraint on economic growth. The economic consequences of diminishing marginal agricultural yield were well understood. But there was no widely accepted suggestion of limits to growth on a planetary scale.

Englishman Thomas Malthus was the 6th of 7 children. He penned a book on population dynamics in 302 BF to counter his father's optimism on the abiding progress of man to overcome Nature's obstacles and engender an enlightened society. "The power of population is indefinitely greater than the power in the earth to produce subsistence for man," Malthus wrote. "Population, when unchecked, increases in a geometrical ratio. Nature herself intervenes."

In his time – and until the end times were upon them – Malthus was a lonely raving pessimist. Industrialization put paid the doubts Malthus raised until the sins of industry overwhelmed Nature herself.

Long-derided Malthus got it right. Humans were inclined to prodigiously breed. Population growth was checked only by the miseries of poverty and famine.

Drunk on their technologies, modern men forgot how tenuous their lives were. World population exploded in the last 2 centuries, raging unchecked despite the obvious danger.

In 300 BF there were 1 billion Earthers. That rose to 1.65 billion by 200 BF. A century later there were 6 billion. World

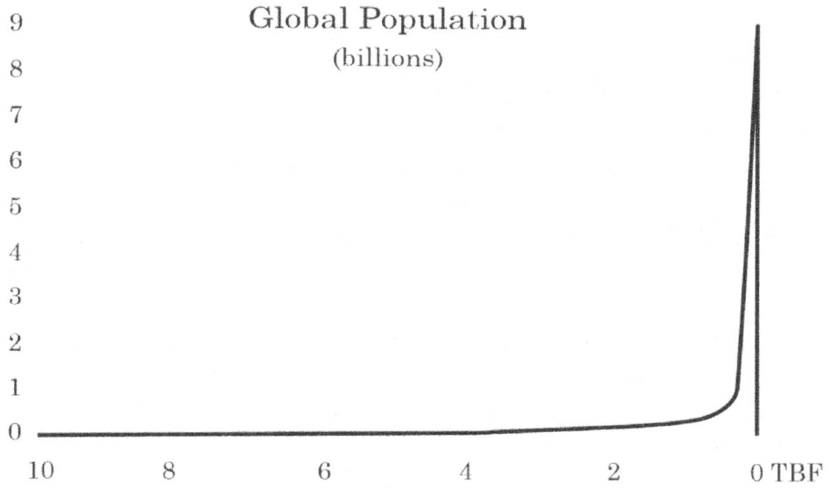


population peaked a half-century before the fall, at 9 billion, before dropping precipitously.

From the late 2nd century BF, educated urban dwellers in some post-industrial societies voluntarily curtailed their breeding. Their professional careers both absorbed and exhausted them. This exception was statistically insignificant. Most went on indulging their biological urges.

The government of China made a brief stab at limiting its population growth. That experiment only lasted from 110 to 85 BF. It was a fiasco.

The biggest problems China faced in trying to limit procreation were mass infanticide and abandonment of girls, as boys were preferred. This caused a sex ratio of 20% more males than females in that generation. 2 decades later, men in want of women produced a rash of sex crimes in China.

On 11 July 113 BF, the UN noted "Five Billion Day": the day that global population supposedly hit 5 billion. The UN changed that day's designation to "World Population Day" 3 years later, to "enhance awareness of population issues, including to the environment." World population had already added another 300 million by 110 BF.

The UN insensibly changed its tune in the 3rd decade of the last century. In 78 BF, anticipating 8 billion in the world that year, the UN called those who fretted about "disastrous overpopulation" "alarmists."

¤✧¤

What may seem obvious in hindsight often appears murky looking forward. People were naturally optimistic. Adversity was internally countered with hope.

When things went awry, there was an inclination to think that life could return as 'normal,' however normal was perceived. People's minds preferred stable patterns. Absent a change in routine, the mind only reluctantly shifted what normal meant.

Regarding self-extinction, the mind's desire to feel secure had people disregarding warning signs. Until the evidence became overwhelming, most folk rejected the notion that the end times were upon them.

¤✧¤

What made Earthers burgeoning population unsustainable was the ravenous consumption that went with it.

Consumption began with resource extraction. 75% of what was taken by moderns was used only once. Among those materials were minerals (50%), ores (for metals) (10%), and fossil fuels (15%). The other 25% of material extraction was biomass. This comprised crops for food, clothing, and other fabric materials, and wood for construction and tools, including paper.

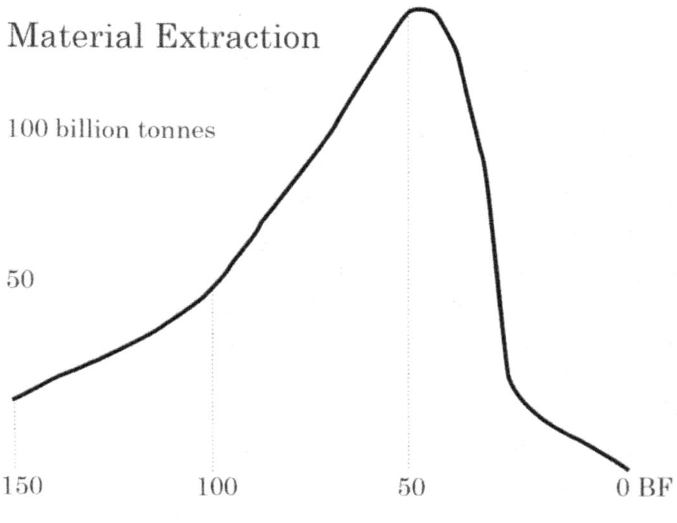

Extraction leapt 6-fold from 150 to 50 BF. Material consumption rose faster than population growth in the half-century before its peak. This was primarily a 'wealth' effect in affluent post-industrial societies.

The rise in material prosperity in poorer countries was nowhere near that of richer nations. In terms of material extraction, poorer societies subsidized the rich.

Man's presence was far-reaching. At the dawn of the 8th century BF, before the Age of Discovery, 6/7ths of ice-free terrestrial biomes were still Nature's domain. By the 1st century BF, only 1/27th was still wild. Men markedly reduced that margin even further in their last century.

¤✧¤

Earthers chronically fouled their planet. Industrialization merely upped the pace of pollution to a horrendous rate. Neither the soils nor seas nor the air were spared.

At the start of the 1st century BF, air pollution was killing 9 million people every year. 1 in 9 deaths were attributed to foul air. Billions of lives were shortened from breathing dirty air. As with other ills, subordinates suffered disproportionately. The toll from bad air rose in the last century as heatwaves flared and wildfires spread across woodlands worldwide.

From the 2nd century BF, numerous national laws, and a much sparser number of international treaties, were enacted to limit pollution. They had some effect, but never enough to staunch the relentless flow of industrial bads.

Along with the final spasm of extraction came a proliferation of rubbish. This dumping of disrespect for Nature littered landscapes for many millennia after Earthers had gone. Landfills and the endless ribbons of roads gave lasting testament to a wasteful race.

The level of trash in 100 BF was 10 times what it had been a century before. In 50 BF, the generation of garbage was twice what it had been just a half-century earlier.

This profusion of rubbish came from embracing the throwaway lifestyle which industrial capitalism thrived on. Unsurprisingly, the level of per capita waste in the richest nations was double that of poorer ones. The most profligate trash makers were those in the military. The protectors of the state generated more than twice as much garbage as the average person.

In the last century, a handful of influential activists in a few rich countries became concerned about the proliferation of rubbish. To avert systemic change, authorities instituted recycling programs for metals, glass, paper, and plastic. It was a ruse. Only a tiny fraction of trash was ever recycled into new products. But the deception worked to mollify public concern.

Being able to propel objects into the heavens gave Earthers' another outpost of refuse. The legacy of the space age was to turn Earth's exosphere into a rubbish tip.

In the 2nd decade of the 1st century BF, as the downhill slide to extinction steepened, space travel was privatized. Outer space had before been a domain exclusive to governments: albeit abetting private enterprise, such as launching commercial communication satellites.

The last-century indulgence of rich men rocketing into space was the ultimate in unsustainable consumption. It was celebrated in institutional mass media as a triumph of the human spirit. Meanwhile, the orbital band around Earth became so cluttered that spacemen were reluctant to go outside their capsules, fearing they might get clobbered by a piece of trash hurtling by.

Trash was not just physical. It was also sonic. Engines had a ubiquitous bad: noise. Whether vehicles on roads, planes in the skies, or ships at sea, engines revved so loud that it lessened the lives of other animals, making them shout to be heard by conspecifics. Even plants were downed by the din of industrialized men.

❀❀❀

As Malthus aptly noted, the controlling variable to Earther's existence was food. The bounty of sustenance set the boundary for all else that occurred.

From the onset of agriculture, crop yields stayed low until the 3rd century BF. Much of what was long believed about soil and nutrition was wrong.

Growing populations were fed only by increasing land under cultivation. As farmers did not understand their ground, continuous cultivation of the same area lowered soil fertility.

Historically, the best farmlands were foreclosed by settlements. Into the modern era, the most fertile soil sprouted only cities, poisonously laced with roads. This disposal of potential was a universal truism. The flattest acreage with a good water supply was the prime choice for putting up buildings.

Japan was exemplary. A mountainous island chain with only 6% of its land arable, Tokyo and surrounding cities sprawled over the Kantō plain, once the country's most productive cropland: so too the Osaka Plain, home to the sprawling cities of Osaka and Kyōto.

⌑✧⌑

Abetting tillage to an unprecedented degree, the heavy plow was invented in the 16th century BF. This innovation spread across Europe over the next 4 centuries.

The benefit of metal farming implements was alloyed. Iron plows polluted the soil and the food that was harvested using such tools. Heavy-metal pollution in the South China Sea from inland farming was significant as early as 4 thousand years before the fall.

Agricultural exchanges in medieval times between Christians and Muslims were fruitful, both in new crops and farming techniques. Arabs introduced the Europeans to the benefits of slave labor for cultivation on plantations.

This model of forced labor was carried forward to the New World, affording prosperity to dominants, especially in the American antebellum South. The economic power of the USA was built upon slavery. Even those states which did not practice slavery benefited by low-cost crops. For instance, prior to the civil war in the mid-3rd century BF, textile mills in the North profited from cotton picked by slaves in the South.

Sugar was a strong stimulus for slavery. Sugarcane was native to New Guinea, where it was first domesticated. The sweetness of this grass ensured its spread: first in southeast Asia and India, and then to amenable tropical regions throughout the world. Europeans took to sugar in the Middle Ages.

Sugar is a tricky crop. Cane must be cut when it is ripe and processed immediately, else it spoils. Hence sugar production was always local to where the crop was grown.

Sugar work was complicated, laborious, and labor-intensive. The need for machinery and an orchestrated workforce for sugar production led some Earther scholars to consider sugar as the starter crop for the labor practices common during industrialization.

Europeans brought sugarcane and slavery with them to the New World. The Europeans had no intention of torturing themselves to produce this beloved food, which could be quite profitably shipped back to the homeland.

Indigenous workers in the Americas were dying of disease and maltreatment. Nothing to do for it but import slaves.

Africa was the closest continent. And it had an abundant supply of enslaved on its coast. As such, Africans were shipped to the Americas in vast numbers. Only 1/3rd of those 'imports' finished the 2-month voyage in the filthy ship bilges they were packed into. Those that survived toiled their lives away as slaves, as did their unfortunate offspring.

In the 3rd century BF, Cuba became the richest land in the Caribbean. Whereas other Caribbean islands were mountainous, most of Cuba formed a rolling plain: ideal for agriculture. Sugar was Cuba's dominant crop.

Unsurprisingly, Cuba retained slavery longer than most Caribbean islands. Sustaining such oppression shaped the destiny of Cuba's political economy. In the mid-2nd century BF, a socialist visionary swept away the rampant corruption there.

¤✧¤

Until 3 centuries before the fall, in every part of the world, commoners lived on the edge of starvation. Mortality from malnutrition remained high. Some part the world hosted a famine every year.

The political economy chosen was behind the chronic hunger throughout history. For instance, the Dark Ages of Europe was sustained by *feudalism*: aristocracy from autocratic land grants for society's upper class. "Feudalism prioritized the needs and desires of lords over the survival of serfs and peasants," wrote 1st-century-BF food journalist Mark Bittman. Those who grew the food starved while their masters grew fat.

Famines were frequent in medieval Europe from the 17th century BF to the 11th century BF. The bubonic plague that struck Europe in the 8th century BF gave relief in food supply, but only because the plague reduced Europe's population by 40%.

For most everyone until 2 centuries before the fall, food meant vegetables. Meat was rare as regular fare except for the upper class.

Meat as a meal for the masses meant massive herds of livestock. Though a modest contributor to self-extinction, meat-eating on the scale it was indulged in the last age was unsustainable.

Similarly, until the 2nd century BF, only near rivers, lakes, or on the coast was fish on the menu. Then, even by the 5th century BF, signs of overfishing were apparent. That depreciation accelerated as fishing fleets were industrialized. The seas were depleted before marine heat did the rest in.

Until the age of machines, most people labored in agriculture. In 250 BF, at the onset of industrialization, over half the world worked the land. The number of those involved in farming only declined as it became more mechanized.

Technology did not change the acreage equation of food production. Increases in agricultural output owed largely to plowing more land. Improved cultivars in the last 2 centuries did improve yields. Nonetheless, crop acreage grew, as did waste.

Earthers grew 95% of their food in topsoil, making soil the foundation of humanity's food system. "Fertility of the soil is the future of civilization," noted 2nd-century-BF agriculturist Albert Howard. Yet the fragility of soil was seldom appreciated. Instead, dirt was taken for granted.

It was long assumed that soil comprises inherently stable chemical compounds. Instead, the organic matter which is critical to soil quality is an active process.

Soil is itself a complex and fragile ecosystem. The soil creation process is so slow that soil is a nonrenewable resource.

Man's neglect of soil often led to its critical vitality being lost. In the last 150 years of human existence, nearly half of the land desired for cultivation was unfit due to pollution and erosion.

Given the dirty treatment of dirt, what mattered most to maintaining soil fertility was fertilizer. Excrement did the job until the 2nd century BF.

The supply of ordure was largely tied to livestock. Livestock supply was limited by fodder crops. At best, the boon of manure barely compensated for having to feed the livestock which provided it.

In China and Japan, farms near cities used "night soil" (human stool), collected by efficient cleaning organizations. In the 280s BF, the first boost in European agricultural productivity came from the importation of guano (bat droppings) from huge deposits off the Pacific coast of Latin America.

The first artificial fertilizer was concocted in 258 BF by English farmer John Lawes, who applied sulfuric acid to phosphate rock. With scant supply of suitable rock, Europeans imported phosphate from Florida, Morocco, and Thailand. Russia used slave labor to mine the rock. The Australians wrecked the landscapes of Nauru and Ocean Island for phosphate.

German chemists developed a process to make artificial fertilizer on an industrial scale in the mid-180s BF. But synthetic fertilizer was not widespread until the 2nd half of the 2nd century BF, when the petrol required to produce it became cheap.

Modern fertilizer use was stunningly inapt. Most of it was wasted. At least half ran off the soil into water courses, where it wreaked environmental havoc.

That did not deter its application. Fertilizer use in western Europe rose 10-fold during the 2nd century BF, but crop production only doubled: mostly by using more land.

¤✧¤

At the turn of the 2nd century BF, the world's geopolitical powers were in Europe and the USA. The 1st World War devastated Europe. The USA was untouched by the carnage, and so became the dominant world power.

The 2nd World War again left the USA unscathed, furthering that nation's preeminence. Food played a significant part in 2nd-century-BF American geopolitics.

The 2nd World War put the capitalist debacle of the Great Depression in the rear-view mirror. Post-war, corporations tightened their grip on the machinery of the USA state. To Americans' great detriment, authorities extensively subsidized the major corporations of all industries. Among them was agriculture.

In 250 BF, half of Americans lived and worked in farming communities. A century later that percentage was less than 10%. That transition happened as corporations turned American agriculture into an industrial production, abetted by governmental largesse.

The *Green Revolution* was an intensification of agricultural conglomeration. The Revolution occurred between the late 160s BF and late 130s BF.

The initiative began by the USA introducing new strains of wheat and rice to economically developing countries. "The Green Revolution was promoted as the solution to world hunger," wrote geologist Dale Allen Pfeiffer. That promotion was pleasant misdirection. The Green Revolution's real aim was to extend the USA's geopolitical influence, including that of American corporations.

The Green Revolution spread agricultural technologies that were known but not widely implemented outside industrialized nations. These included higher-yield cultivars, modern irrigation techniques, synthetic fertilizer, and biocides. The idea was to turn farms into factories, with plants as the slave labor – though indigent farm workers also figured in.

Prior to the Green Revolution, many developing countries struggled to feed their burgeoning populations. India was exemplary. Famines were frequent in India from the 150s BF to the 120s BF. Most of the population starved or were malnourished. The Indian government fostered this disaster by not providing financing or adequate compensation to family farmers for their labors.

In terms of sheer productivity, the Green Revolution was a success. Cereal production more than doubled 140–115 BF.

But the externalities of the Green Revolution were extensive. Finer farming through chemistry proved a chimera. The short-lived productivity boon of the Green Revolution was spent as its ersatz alchemy exhausted the soil. In the end, the Green Revolution became another example of political false advertising.

By the end of the 2nd century BF, nearly 40% of Earth's land surface was given over to agriculture. That figure rose further in the 1st century BF.

The folly that fed humanity emitted 1/3rd of man-made warming emissions. "Expansion of land area used for agriculture is a leading cause of biodiversity loss and greenhouse gas emissions, particularly in the tropics," explained ecologist Ben Phalan in the early 1st century BF.

As Malthus predicted, the greater abundance of food in the developing world provoked a burgeoning birth rate. World population increased 32% in the quarter century of 140–115 BF: from 3.3 to 4.85 billion.

¤✧¤

Food quality declined in the last century and a half before the fall. This degradation had 2 causes: the industrialization of food and consumer choice.

1st, foods became less nutritious. This downshift owed to selection and industrial processing.

Phytonutrients were the vast array of compounds – over 25,000 – that conferred health benefits in their consumption. "Eating fruits and vegetables without phytochemicals is like drinking the empty calories of a can of soda," observed 1st-century-BF nutritionist Jed Fahey. Alas, phytochemicals typically had a bitter taste, so growers selected varieties that were sweeter but less nutritious.

Grapefruit exemplify. White grapefruit is high in naringin, a bitter phytonutrient with anti-inflammatory, anti-ulcer, and anti-cancer properties. Pink & red grapefruit achieve their sweetness in large part by having substantially less naringin.

Florida was the grapefruit capital of North America. In 115 BF, Florida sold 27 million boxes of white grapefruit, and 23 million boxes of the colored varieties. 30 years later, Florida growers shipped twice as many colored grapefruit as they did white ones. 3 decades later, white grapefruit were scarce in produce markets.

The seeds called "grains" were fruits. Grains were some of the most widely consumed foods. Though considered a grain, corn was really a vegetable.

In much of the modern world, wheat was the grain most gobbled. Wheat exemplified how Earthers processed the nutrition out their food.

The *bran* of a wheat berry is its fibrous outer layer. Eating bran was healthy exercise for the microbes that performed human digestion. The most nutrient-rich kernel was termed the *germ*.

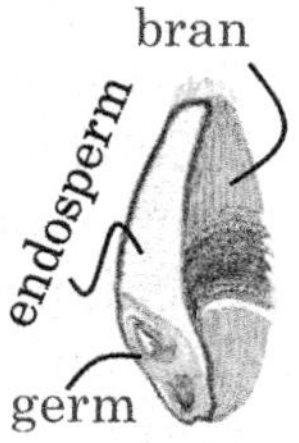

wheat berry

Taking away the bran and germ leaves wheat a starchy *endosperm*. The endosperm comprises the bulk of the berry: carbohydrates that don't deliver much nutrition.

The nutrients in the bran and germ are embedded in oil. That oil can go rancid, ruining the flour by which humans ate wheat. Hence, flour from whole wheat was viable for only a limited time.

Turning wheat into a commodity with a long shelf life required removing most of its nutritional value. Which is exactly what happened.

Deception was leavened into modern commercial bread. The empty calories of white bread were marketed as "building strong bodies 12 ways." All 12 ways were from synthetic supplements.

Refining rice underwent a similar process: immortalizing the food by robbing it of its nutritional value.

The 2nd way that foods became less nutritious came from what people chose to eat. When they could afford to, people ate more meat and milk products. Neither food was good for the body.

Milk is a mammal's first food. Mother's milk is the natural diet of infants. For children, breast feeding conferred health benefits that lasted a lifetime.

Among mammals, only humans drank milk past infancy. The common milk that people consumed came from cows. Cow's milk had 3 times the protein of mother's milk. Like all animal protein, milk metabolizes acidically. Calcium, an excellent deacidifier, was drawn from bones to neutralize cow milk's acidifying effect.

The result of swilling milk was disrupting the digestive system and robbing the body of calcium. Yet cow's milk was advertised, with government approval, as being calcium-rich and highly nutritious. The deceit derived from touting the chemical ingredients of milk while ignoring how milk was digested and affected the human body.

Moderns consumed more industrially processed foods and fewer fresh fruits and vegetables. Created for addictive taste appeal, processed foods had excessive sugar, salt, empty calories, and nutritionally dubious additives. "The food industry became the junk-food industry," remarked Mark Bittman.

People were excessively fond of sweets. Refined sugar was a major culprit in creating chronic health problems.

To keep their profits sweet, the food industry ladled ever more sugar into their products as the 1st century wore on. They did so because the taste buds of consumers, as their health deteriorated, were desensitized to sweetness.

Processed foods also had a higher environmental impact than nutritious ones. Processing foods to lower nutritional value wasted water, energy, and materials, especially in its packaging. The truly modern Earther food was wrapped in single-use plastic.

Evidenced by their dietary habits, most moderns appeared to know little about nutrition, nor cared to learn. Obesity soared worldwide in the last century, as did chronic diseases from self-induced malnutrition.

¤✧¤

The Green Revolution's sole focus was the business of food supply, which was only half of the hunger equation. Food security – people's access to food – stayed shaky for neglected subordinate tribes throughout the world.

There was more than enough food for everyone. The problem was distribution. The quarter of the world's population that lived in industrialized countries ate half the food. Even in those countries, hunger was commonplace.

The poorest countries exported their food to the richest. Plutocratic control guaranteed chronic food shortages and bouts of famine in poor nations.

This was nothing new. Selective undernourishment had been on the political agenda since history began. "Hunger has always been political," remarked 1st-century-BF agricultural ecologist Gordon Conway.

In the USA – one of the wealthiest nations – 15% did not get enough to eat, including 25% of the children there. Those who went hungry were in subordinate tribes. Dominants threw out half of the food they had foraged in supermarkets.

"We have the means to wipe hunger and poverty from the face of the Earth," proclaimed American ruler John Kennedy in 137 BF. Those means were never applied. Aside from measly handouts, governments were indifferent to those in need.

In the early 1st century BF, over 1/3 of the food produced globally was wasted. Meanwhile, 1/3rd of the world's population went undernourished.

Despite the widespread scarcity of sustenance, food was typically treated as a disposable commodity. That cavalier attitude heightened noticeably in wealthier nations toward the end of the 2nd century BF. "In one day, a supermarket can easily throw out enough to feed over a hundred people. Supermarkets will deliberately overstock because they believe that shoppers like to see full shelves, which give the illusion of infinite abundance," commented 1st-century-BF ecologist Tristram Stuart.

People in rich nations were egregiously wasteful with their food supplies. Britons threw away 720 million edible eggs in 82 BF. "In poor countries, food loss is primarily due to the lack of storage and transport; while in wealthy nations, food waste is a result of profligacy and inefficiencies toward the end of the food supply chain," the UN reported that year.

Because of the inefficiency inherent in the disorganized market system, the massive level of food waste equaled the production acreage of land under cultivation. Half of the food grown was not eaten.

The bounty of the seas was also wasted. Even as fish stocks plummeted, ships threw away seafood that was not what they were fishing for. Chucking away catches of edible species was routine. Sardines were not appreciated in some European countries, and so were discarded. Meanwhile, fleets from other countries went out to catch sardines.

Governments ham-handedly attempted to limit overfishing. The result was greater waste. If a vessel hauled in more than its quota, it threw back the rest. 75% of the fish thrown back perished. All told, at least half of what was caught was jettisoned.

Shrimp was an extremely popular seafood. For every kilogram of shrimp caught, over 10 kilos of marine life died.

Like other food waste, blemished fish were discarded. 30% of the fish caught never made it to anyone's plate.

Less than a decade after Mao Zedong seized power, China suffered a horrendous famine from his erratic governance. Some 50 million Chinese starved between 142 and 138 BF.

Xi Jinping survived those distressing days. When he became China's commander, Xi instituted a web of edicts to eat "in a civilized and healthy way." China was the only nation to have a food waste policy.

Food prices in much of the world jumped dramatically in 93 BF. The extensive hunger sparked riots in over a dozen countries. A few regimes in Muslim nations were toppled in subsequent years from civil unrest initiated by the food crisis. That episode of widespread food riots was a harbinger of the decades that followed.

Russia invaded neighboring Ukraine in 78 BF. Both countries were significant grain exporters, including to poorer nations in Africa and Asia. World food prices markedly leapt that year, mostly from national hoarding. Such jumps in food prices became regular events as the fall approached.

Crop yields fell in the last century from a confluence of causes, including heat, extreme weather events, and soil exhaustion. Financial speculation worsened the situation. World food stockpiles diminished.

People eating more meat aggravated the problem of food supply. By the 2nd decade of the last century, livestock were eating more grain than people.

The global food situation sorely deteriorated in the half-century before the fall, as international trade broke down. Famine was an unceasing plague, especially in the poorer countries where hunger had been a chronic problem. In 78 BF, David Beasley, head of the UN food program, foresaw "global destabilization, starvation, and mass migration on an unprecedented scale." Yet no concerted campaign was conjured to avert that fate.

Starvation took a severe toll in the last century. As it had been during much of their existence, hunger was a leading cause of death during Earther demise.

¤✧¤

Man labeled as noxious those life forms which were imagined as lessening crop yield. Plants not intended which sprouted alongside those sown were *weeds*. Insects and other creatures considered as downing plant productivity were damned as *pests*.

To counter these conceived threats, men employed *biocides*: chemicals meant to selectively deal death to small animals (pesticides) and weeds (herbicides). Biocides had been applied since agriculture took root.

Biocides were invariably bads: fouling the soil and taking lives beyond weeds and pests. Biocides were selectively banned only after their death toll had been proved severe.

The plying of pesticides started in antiquity. Farmers in Sumer dusted sulfur on their Mesopotamian crops 5 millennia before the fall. Contemporaneous Indian texts mention using poisonous plants for pest control.

Ancient Romans slayed insects by burning sulfur. Weeds were treated with salt.

By the 7th century BF, toxins such as arsenic, mercury, and lead were applied to crops to kill pests. In the 400s BF, ants were fed a mixture of honey and arsenic.

Organic pesticides were also employed. Beyond their ample use as produce, plants were long treasured as both medicine and toxin.

Pyrethrum, derived from chrysanthemum flowers, was used as an insecticide from 2500 BF. In the Middle East, pyrethrum long served as a lice remedy. The locals called it *Persian powder*.

The rise of modern pesticides began during the 2nd World War, when a variety of killer formulas found favor. Among them was DDT. These synthetic biocides were products of USA government-sponsored research.

Potent pesticides were quickly adapted to civilian use when the guns stopped in 155 BF. The triumphalist rhetoric used to describe the war effort was readily translated into the need to fight and win the war on weeds and pests. An American herbicide advertisement in 154 BF read: "Get rid of weeds – now that chemistry has developed an easy, economic method for their destruction."

These new toxins were inexpensive and enormously popular. Americans were especially enthusiastic, as they were naïve in their love of modern technologies, and in their view of economic growth as "progress." Common was the mistaken assumption that increasing agricultural output was by definition good – and that this bounty was inextricably tied to modern techniques, including the application of biocides.

The popular American publication *Saturday Evening Post* published an article on 23 October 146 BF which read: "Airplanes swoop over a field leaving a trail of vapor, engaged in chemical warfare against crop-destroying pests. You'll also see spray machines squirting selective chemicals which kill weeds but don't harm the crops." That vapor was DDT.

Lulled into a false sense of security about "better living through chemistry," people liberally applied pesticides to "sterilize" their habitat of pests. In the 3 decades following World War 2, pesticide use in the USA leapt 10-fold.

DDT was a colorless, tasteless, and almost odorless biocide. DDT was first synthesized in 226 BF. Its insecticidal activity was discovered in 161 BF.

The American military extensively sprayed DDT in Europe and the Far East during World War 2. They marveled at its lethal potency. When the war ended, DDT garnered vast application, celebrated as "the atomic bomb of the insect world." The praise was on-point. DDT and other synthetic pesticides wreaked their havoc by ripping apart atoms. "The damage caused by the ionization of atomic fallout and by chemical agents are the same," noted 2nd-century-BF chemist Americo Mosca.

2nd-century-BF biologist Rachel Carson published a seminal book on biocides in 138 BF, titled *Silent Spring*. "Since the mid-1940s [150s BF], over 200 basic chemicals have been created for use in killing organisms descried as "pests." These sprays, dusts, and aerosols are now applied almost universally to farms, gardens, forests, and homes – nonselective chemicals that have the power to kill every insect, the "good" and the "bad," to still the song of birds and the leaping of fish in the streams, to coat the leaves with a deadly film, and to linger on in soil – all this though the intended target may be only a few weeds or insects. Can anyone believe it is possible to lay down such a barrage of poisons on the surface of the Earth without making it unfit for all life?"

Carson targeted DDT as an especial threat. DDT was readily absorbed into soils, where it actively exposed itself to organisms for decades. DDT worked its way up the food chain, thus incrementally adding to its toxicity on larger predatory animals. DDT was carcinogenic to humans.

The USA government's agricultural agency fought against forbidding DDT. But Carson's exposé of DDT's death-dealing made a public impression that politicians dared not ignore. DDT was banned in the USA in 128 BF, a decade after *Silent Spring* was published.

DDT continued to be used in Africa and South America to thwart insect-borne diseases, such as malaria. Some countries, notably India, continued to apply DDT to crops until their agriculture utterly failed.

Chemists in the 5th century BF extracted nicotine sulfate from tobacco leaves as an insecticide. But the lethality of nicotine only found copious appreciation from the 110s BF, when pesticides known as *neonicotinoids* were profusely employed. Neonicotinoids were among the 2nd wave of synthetic biocides.

Neonicotinoids worked by mimicking neurotransmitters: putting nerve cells into terminal overdrive. Though especially deadly to insects in minute quantities, this mechanism debilitated all animals.

Like DDT, neonicotinoids unleashed an insidious pandemic of biocidal damage. Pollinators were pummeled. The loss of pollinating insects by the early 1st century BF abetted plummeting crop yields. Yet neonicotinoids kept being applied in increasing quantities.

A significant fraction of the temporary gains in harvests during the Green Revolution were attributed to biocides. This encouraged their extensive use.

By the early 80s BF, over 4 million tonnes of biocides were applied annually. China was one of the most intensive users, as was Japan.

In fact, biocides made scant difference in crop productivity: only an estimated 8% at most. Meanwhile, "pests" and "weeds" adapted to overcome the chronic toxins.

The impact of pesticides was incalculable. They begat the mass extinction of insects. As insects were critical to the viability of all terrestrial biomes, their diminution was a harbinger.

By the end of the 2nd century BF, the post-war chemical revolution was an entrenched institution. Synthetics of all sorts pervaded modern life. Most were produced using petroleum.

The externalities of chemical syntheses were crippling. Autism, retardation, and other developmental diseases soared in the last century. This diminishment owed to various pollutants, including biocides just to grow food.

⌑✧⌑

The furless apes of Earth still had pelts on their heads and pubic parts. Especially for women, having attractive hair was an aesthetic concern for over a hundred thousand years. Early combs were made from animal bones and wood.

Manufactured from the 220s BF, celluloid was the 1st man-made plastic. Comb makers were early converts. From then, the common comb was a plastic product.

The word *plastic* derived from the ancient Greek verb *plassein*: to mold or shape. Plastics had that facility thanks to their molecular structure as a robust carbon polymer. Long, flexible molecular components (monomers) repetitively bond into a polymer. These polymers may be fiddled as desired. A plastic may be flimsy or stiff, depending upon its chemical composition and thickness.

Synthetic plastic replaced many natural plastics: some just before the source expired. Among natural plastics were the tusks and shells of animals.

Combs illustrate the dilemma. Combs were made from cattle horns for centuries. But ranchers stopped bothering to dehorn their bulls. Combs made from ivory were exquisite. Ivory demand drove elephants to the brink of extinction. Likewise, the shell of the hawksbill turtle had been another favorite for combs. The hawksbill too was taken to the brink.

Looming shortages in natural sources spurred the quest for synthetic plastic. Invented in 193 BF, Bakelite was the 1st man-made plastic. Bakelite was conceived to substitute for shellac, which had been coming from the sticky excretions of the female lac beetle. It took 15,000 beetles 6 months to produce a pound of shellac.

Electricity as a power source started in the early 2nd century BF. The contemporaneous discovery that shellac made an excellent insulator for electrical wires saw shellac's demand skyrocket.

Plant cellulose was the raw material for the earliest plastics, including celluloid. But petroleum and natural gas proved a cheaper source of the hydrocarbons that comprise plastic. So it was for Bakelite and beyond.

The extensive use of petrol plastics began in the 150s BF. Plastic was precious because it could be fabricated in infinite ways: from films to fabrics to sturdy solids.

Since the invention of bronze 6,000 years earlier, no other substance had as much impact on material production as plastic. The comparison between bronze and plastic was ecologically telling. Bronze had a much longer life and was recyclable.

The thought of recycling plastic did not even enter into the minds of those enthralled by this prodigious wonder. In

145 BF, the popular American magazine *Life* celebrated the dawn of "throwaway living" thanks to disposable plastics.

Most of the plastic produced was used only once before being thrown away. Only a minuscule fraction of all plastic produced was recycled.

Plastic production went from less than 1 million tonnes in 155 BF to 600 million tonnes in less than a century.

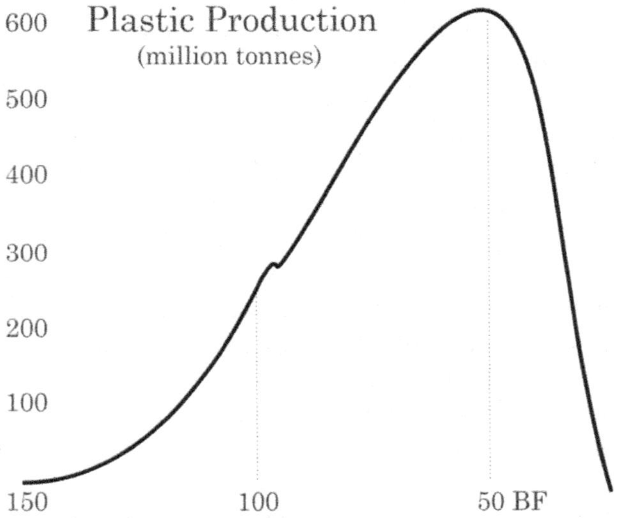


Unsurprisingly, plastic pollution became a severe global problem. Rivers, canals, and beaches became clogged with plastic debris. Massive plastic deposits formed in the world's oceans, corralled by currents. Soils were contaminated with plastic, affecting crop production.

Countless billions of creatures died from plastic. Marine animals were snared in discarded fishing nets. Degrading plastic had an alluring scent, resulting in fatal indigestion. Plastic underwent physical and chemical transformations as it broke down, increasing toxicity.

Plastic waste permeated the most remote places on Earth, including the deepest ocean trenches and polar caps. By the early 1st century BF, degraded plastic – microplastics – permeated the drinking water and food people ate. Humans contaminated their own blood and cells with plastic.

For all the damage done by plastic, it was a fraction of the rubbish problem. Only 10% of the solid waste discarded was plastic: such was the scope of trash capitalism produced.

No effectual countermeasures were taken to plastic pollution. "A lot of these chemicals and materials are necessary for our lives now," noted ecologist Carney Almroth in 78 BF. For its convenience, plastic killed billions of creatures.

❀❀❀

From prehistoric times, men routinely killed animals as "game." That term – *game* – indicates the joy men felt killing for sport. As larger animals were harder to "bring down," they were especially prized. Trade begat slaughter for profit.

Hominids roamed through Africa and Eurasia for millions of years. *Homo erectus* and his fellows felled large numbers of large animals with the flint implements they had crafted. Mammoths, elephants, elk, hyenas, and bears were among them. The lion lasted in Europe until classical times.

Over 90% of the sizable animal species which inhabited Australia went extinct 40 millennia before the fall, after the arrival of men who hunted them down. Several of these were large marsupials unique to that continent.

The mastodon was a huge, hirsute beast with a trunk and fearsome tusks. Mastodons lived in the Americas for millions of years. Men encountered mastodons after they arrived on the continent 26,000 years before the fall.

15,000 years later, the mastodon had been hunted to the brink of extinction. A flash freeze finished them off.

The mastodon was but 1 of 35 genera of large American mammals which went extinct in this epoch. Other doomed megafauna included saber-toothed cats, large llamas, massive muskox, and giant ground sloths.

⌑✧⌑

Mammals stayed small for 10 million years after the mass extinction event that downed the dinosaurs. This gave birds, the direct descendants of dinosaurs, a window of opportunity.

Small herbivorous birds flew to occupy the islands of Madagascar (from Africa), Mauritius (from India) and New Zealand (from Australia). Once there they upsized, becoming too big to be overtaken by the local small predators. Gaining girth weight put getting off the ground out of reach. Being earthbound made them easy pickings when men arrived.

Madagascar was a large tropical island that separated from the landmass which became India 88 MBF. While Madagascar lingered off the east coast of Africa, the Indian subcontinent continued to the northeast, crashing into Asia. Madagascar's long isolation allowed a wide variety of unique animals and plants to evolve.

Men arrived on Madagascar 1200 BF. They slaughtered the animals which they did not domesticate into livestock. Elephant birds were among those killed off. In the millennium that followed, wanton deforestation drove mass extinction in Madagascar.

Moa in New Zealand suffered the same fate as other flightless birds in the late 9th century BF. 1,500 settlers managed to wipe the moa out in short order.

The Dutch took possession of Mauritius, an Indian Ocean island, in 502 BF. Within 64 years, men had extinguished a fearless, flightless bird, from which came the cliché: "dead as a dodo."

Early European settlers in North America frequently commented on the countless number of graceful blue pigeons that filled the skies. "There are wild pigeons in winter beyond number or imagination, so thick that they have even shadowed the sky from us," remarked one the first settlers in Virginia.

Apart from hawks and eagles, passenger pigeons lacked natural predators. These birds were omnivores: eating berries, seeds of all sorts, worms, and small insects. The bounty of Nature let passenger pigeon populations grow to vast numbers.

These pigeons were surprisingly vulnerable to humans. As a female laid only 1 egg a year, losses were not readily replaced. Nesting in immense colonies and migrating in huge flocks rendered the birds easy to attack by murderous beasts.

Native Americans captured passenger pigeons in large nets. By the 470s BF, European settlers were doing the same.

The scale of slaughter was limited until the early 220s BF, when railroads linked the Great Lakes with New York. Then it took only 50 years – supplying cheap meat to the East Coast – to wipe the passenger pigeon out: a bird that once numbered in the billions.

⌑✧⌑

There was a relentlessness to man's casual carnage of wildlife. In the last century of civilization, many thousands of animal species were listed as "endangered" before slipping into oblivion. The listing was mere homage.

The toll of mass extinction was far beyond the meager reckoning men belatedly noticed. The tallied statistics of slaughter did no justice to the insouciant snuffing of life that occurred on a legendary scale. It was as bad as the Great Dying: the mass extinction event 252 MBF that had previously overshadowed all others.

❀❀❀

The knitting together of economies afforded luxuries which would not otherwise been economical. Albeit unevenly, globalization enriched its participants.

Even in the machine age, labor accounted for 3/4ths or more of the cost of a product. Squeezing out profits by keeping costs low, companies sought to manufacture where labor was cheapest.

Thanks to unionization, workers in the USA had become accustomed to comfortable living in the decades following the 2nd World War. Though extensive poverty still lurked underneath, that country had a sizable middle class, as well as an exceedingly rich upper crust.

American corporations were at the forefront of *offshoring*: relocating labor-intensive activities to a foreign country where workers could be paid a pittance. Offshoring was the economic inverse of slavery: moving to where labor was cheap rather than importing cheap labor.

Neighboring Mexico was initially favored by American corporations in the early 120s BF, when the trend began percolating. Finished products were shipped back into the USA for sale.

Rich European countries followed the USA's lead in offshoring to cut labor costs: thus the knit of modern globalization was sewn.

Toward the end of the 2nd century BF, the industrially developing Far East became a favored offshoring region. By the 1st decade of the 1st century BF, China had become the world's foremost factory. Offshoring crowned China as a great global power in the 7 decades before world civilization collapsed.

Offshoring weakened a nation while strengthening the country that received the work. Workers in the offshoring country lost their jobs and income. Over time, the skills of the labor pool in the offshoring nation withered. This dynamic was influential in the decline of the USA and the rise of China.

❀❀❀

While Earth tends to icehouse, hothouse itself was not a detriment to abundant life. Dinosaurs were most prolific when Earth had no ice on its polar caps.

Mass extinction happens with abruptness, whether by changes in climate or geophysical upheaval. Extensive pollution might also snuff life on a massive scale, as it did in the Earther extinction event.

The doom of dinosaurs occurred when a pair of huge bolides slammed into central America and off the coast of west Africa. Each of these fireballs was around 9 kilometers wide.

The collisions produced the most powerful earthquake ever. Impact shock waves coursed through Earth, catalyzing massive floods of basalt and volcanic eruptions on the other side of the planet.

The thermal result was rapid cooling. Dust and soot darkened the skies for nearly 2 years. 5 years after impact, Earth's atmosphere had chilled 26°C. The soot came from raging wildfires. Underneath the American strike point was a rich reservoir of volatile crude oil, with extensive forests nearby.

Unlike the instant icehouse which did dinosaurs in, humans went out rocketing to hothouse. The dynamics of chemistry meant that humanity's thermal destiny began in the oceans. Water warms more slowly than air. The breath of Gaia is such that atmospheric heat settles into the seas.

The oceans absorbed 93% of the warmth from greenhouse gas emissions. It took 4 to 3 decades for this vast reservoir of marine heat to equilibrate into the atmosphere. The thermal transfer rate sped up as the world got warmer. The upshot is that doom was baked in before common folk took global warming seriously.

Earth has temperature-sensitive ocean and atmospheric currents. These currents seasonally set weather and occasionally delivered extreme events. The global warming that ensued in the wake of industrial pollutants altered those currents to striking effect.

Storms, floods, and droughts worsened because atmospheric currents stagnated, generating greater intensity. Tornados multiplied and strengthened in biomes which had been subject to them, such as the central and southern USA. So too tropical cyclones on the other side of the world.

The polar vortices were atmospheric currents around Earth's poles. When strong, the vortices constrained cold air from straying from the poles. These vortices weakened seasonally before global warming altered their dynamics. As the climate warmed, cold air containment became more erratic.

Global Air Temperature

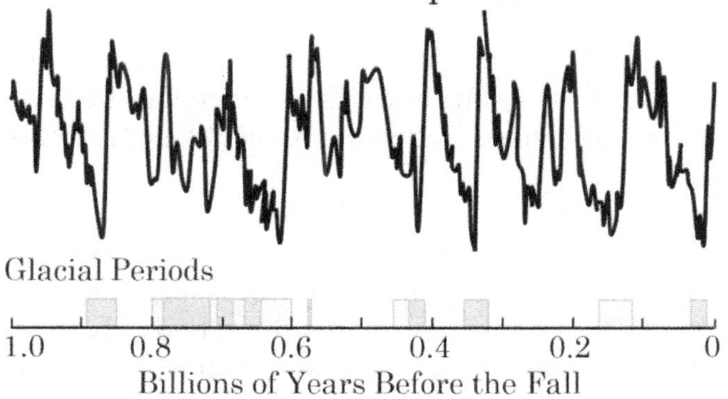

In the early 1st century BF, a sporadically weakened Arctic polar vortex brought frigid winter weather events to the northern hemisphere. Similarly, Australia suffered "polar plunges" when the Antarctic vortex unseasonably lapsed. By the mid-60s BF, these raging cold spells had ceased, as polar frigidity had turned tepid.

Climate change in the 1st century BF had profound pronouncements. Patterns of precipitation intensified as Earth warmed. Biomes tending to aridity dried out whereas those prone to rain were drenched. This water redistribution was calamitous.

¤✧¤

Earth is a watery world. At the start of the 1st century BF, oceans covered 71% of Earth's surface and held 97% of the planet's water. 2% was frozen in glaciers and ice caps.

Less than 1% was freshwater. And, until the last century, only 32% of that water was liquid. The rest was frozen: locked up in glaciers and ice caps. All told, only 1/10,000th of 1% of total water on Earth was accessible by plants and animals. In essence, the supply of freshwater was fixed.

Freshwater was essential for all life on land. Before human overpopulation and industrialization, Earth had an abundance of freshwater. By the 2nd century BF, freshwater had become a precious commodity, as humans were polluting it far faster than its resupply. Still, men wasted water as if there were an endless supply.

By the last century, over 2/3rds of the freshwater bodies around the world had been sorely fouled. Groundwater too was dangerously laced with toxic contamination most everywhere.

Economic development depended upon water. Food production and industrial manufacture relied heavily upon freshwater use. These industries took 80–90% of the water consumed.

As with all resources, waste was a matter of financial wealth. People in industrialized nations used 10 times more water at home than those in lesser-developed countries.

By the fall, in many places, Earthers had exhausted their water supply. Some of the worst water shortages were in the most heavily populated countries: India and China. In the end, the death toll from lack of drinkable water exceeded that of famine. Over half of Earthers died in the last century from nothing to drink or eat.

¤✧¤

California was on the west coast of the USA. At the onset of the 1st century BF, 75% of California's precipitation fell in the northern part of the state. Much of it was winter snow in the Sierra Nevada mountains. Meanwhile, 75% of the water consumed was in central and southern California.

As such, the Californian economy relied upon an extensive system to porter water south. Without this transport, the Central Valley would have been a grassland rather than one of the world's most productive agricultural regions. And southern California would have been too parched to populate.

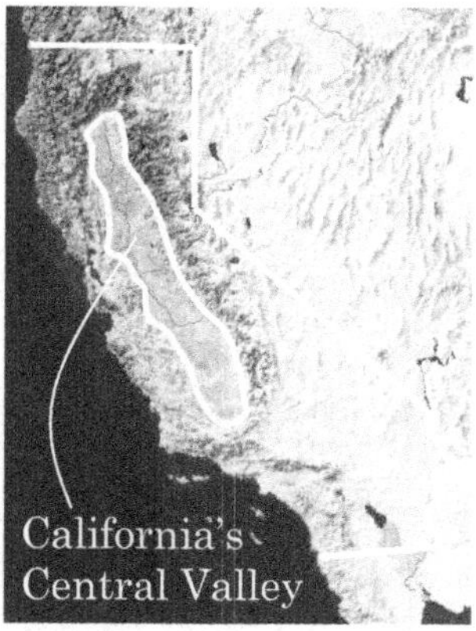


California's Central Valley comprised 13.7% of the state's land mass. Covering less than 1% of the country's cropland, the Central Valley grew 8% of America's food.

That productivity depended upon massive irrigation. In the early 1st century BF, 17% of the irrigated land in the USA was in the Central Valley.

The intensive agriculture heavily contaminated the groundwater with fertilizers, biocides, and animal wastes. Oil extraction added to the befoulment. The supposed potable water in the Central Valley was not safe to drink.

"Drought is the new normal," announced California ruler Jerry Brown in 86 BF. The state had got a goodly portion of its water supply from the nearby Colorado River, but that too was near exhaustion.

The state continued to dry out, as did the Colorado River. By the end of the 70s BF, California was rapidly depopulating.

A selfsame dynamic affected other habitats which had been abundant suppliers of food to millions of people. Among them was the Po Valley in northern Italy.

The Po Valley was the drainage basin of the Po River. The Po was Italy's longest river.

The Po River was fed by snowmelt from the Cottian Alps, which bordered southeast France and northwest Italy. The river flowed east, ending in a delta that dumped into the Adriatic Sea. The city of Venice sat in that delta.

The Po Valley had been inhabited for some 800,000 years. The valley had initially been a refuge from glaciation.

Extensive cultivation in the Po Valley first occurred between the 12th and 9th centuries BF, during a lukewarm climate phase in Europe. Woodland clearance and wetland drainage transformed the natural landscape.

The Po Valley had the advantage of evenly distributed rainfall throughout the year – until the last century. In the 70s BF, drought dried the Po Valley like California's Central Valley.

The Po Valley also had some of the worst air pollution in Europe. "The Po Valley is unhappily situated for atmospheric pollution in terms of climate and geography. Wind is rare and there are frequent and prolonged episodes of climatic inversion. This means that the air is still," explained 1st-century-BF Italian environmentalist Damiano Disimine.

❀❀❀

From the 2nd century BF, marine species were eliminated at twice the rate of land animals. This initially owed to excessive exploitation and egregious pollution. Then extreme ocean warming took its toll.

Like the extravagant wasting of freshwater, Earth's oceans were treated as if they were an infinite toilet. By the 2nd decade of the 1st century BF, sea bathing was no longer safe at the world's beaches.

Ships were always the cheapest, fastest way to move cargo over long distances. Most of the world's populations lived in coastal areas, and on waterways, for this very reason. Shipping had been the means of global exploration and was the lifeblood of economic globalization in the industrial era.

By the early 1st century BF, over 60,000 colossal ships crisscrossed ocean trade routes. Their propellant was the filthiest petrol: so foul that its use on land was outlawed. The 16 largest ocean cargo ships produced as much sulfur pollution as the entire global fleet of cars. Atmospheric sulfur results in acid rain.

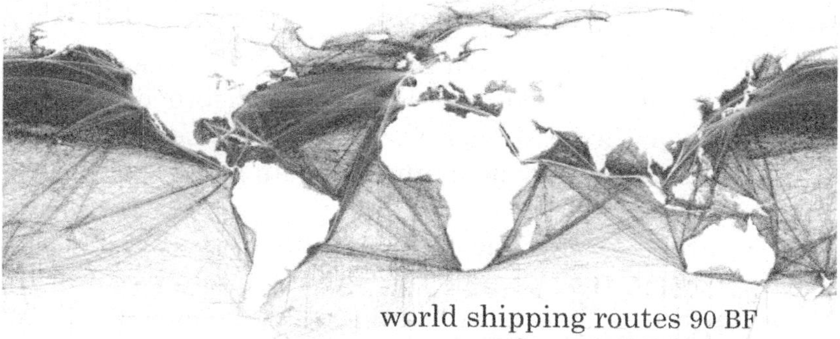

world shipping routes 90 BF

Ship exhaust affected local weather. Chemical agitation intensified thunderstorms above busy shipping lanes.

Large marine mammals were regularly run over by trade ships. The noise of ocean vessels disrupted and shortened the lives of nearby sea creatures.

Much of the world's oil was transported by ship. Oil spills of all sizes were a regular feature of ocean shipping. Those areas polluted by such a spill did not recover for a century or more. Shipping dotted the oceans with death.

¤✧¤

Coral were colonial marine invertebrates. Individual polyps grew together to form colonies by excreting stony exoskeletons which anchored them to the seabed.

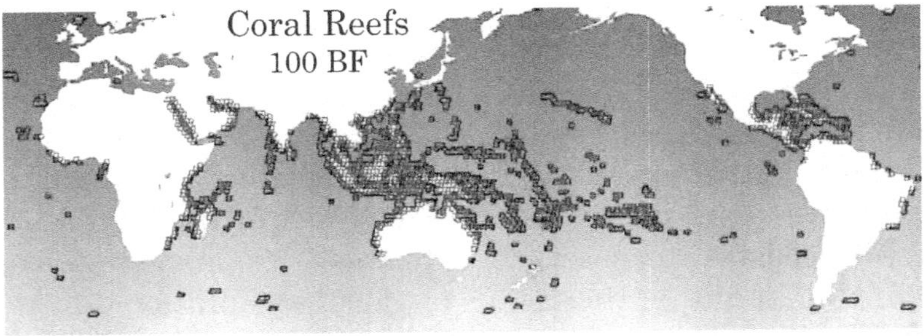


Coral provided the base for reefs which often extended for many kilometers. These reefs were critical buffers preventing shoreline erosion, and thus stabilized their habitat. Further, by their absorption of carbon dioxide, coral reefs helped regulate global temperature.

Coral reefs sat mainly in shallow seas sporting warmish waters. These reefs were sensitive to light, temperature, and water acidity (pH). It took millions of years for Earth's coral reefs to grow to their peak at the onset of the 2nd century BF.

A *keystone species* is one that has a disproportionately large effect on the health of its habitat. Keystone species play a vital role in creating and maintaining an ecological community.

Coral reefs formed one of the most diverse ecosystems on Earth. Though they occupied less than 0.1% of the world's ocean surface at their peak, coral reefs were home to 25% of all marine species. That set coral among the most significant keystones on Earth. Many modern societies relied upon the seafood bounty provided through coral reefs.

The largest coral system was the Great Barrier Reef off the coast of eastern Australia. It was extinct by 72 BF. Ocean warming, acidification, pollution, and deliberate devastation killed all coral reefs less than a decade later.

⌑ ✧ ⌑

Coral reefs were just one facet of marine extinction. Men mercilessly fished the seas as fiercely as their technology allowed. "Until the 20th century [2nd century BF], the stocks of fish in the vast oceans of the world seemed to be inexhaustible. There was no attempt to limit catches. All effort was put into maximizing exploitation," wrote English historian Clive Ponting in the early 1st century BF.

As with other environmentally destructive initiatives in the last epoch, the USA was an early promoter of overfishing. Sensing the political struggle, other countries responded in kind.

Fisheries first began collapsing in the late 3rd century BF. Under American encouragement, this accelerated after the 2nd World War. From the early 1st century BF, latecomer China enthusiastically furthered fishery decimation. Robbing the sea of fishes had a domino effect on sea birds and other life that had relied upon healthy oceans.

By the early 60s BF, the seas were spent. Fishing harvests were a tiny fraction of what they had been a century before, when 1/3rd of the world had already been fished out.

Marine warming, acidification, deoxygenation, and chemical pollution extinguished sea life. The final straw was loss of plankton, which humans had not bothered to monitor. The result was catastrophic. By the fall, Earth's oceans were amid an unprecedented mass extinction event, from which recovery would not begin for nearly a million years.

❀❀❀

The omnivore hominin body never adapted to mammal meat as a healthy food source. Further, meat incurred additional burdens beyond its consumption.

*Zoonosis* was a disease caught from other vertebrates. 60% of the infectious diseases people were subject to derived from eating the meat of land animals. Zoonotic diseases were difficult to control because of their animal reservoirs, viral ones especially so.

Viruses constantly self-evolve in their pursuit of perfecting parasitism. Further, they trade tips on how to infect, become more contagious, and other tradecraft.

Bats had a uniquely impervious immune system. As such, bats were a prolific incubator of zoonotic viruses. Bat's ability to harbor viruses afforded the viruses time to reside and concoct greater contagion before heading out to infect other animals.

A cold was a disease centered in the breathing system. The "common cold" was among the most frequent contagions people suffered. Rhinoviruses and coronaviruses were responsible for the common cold.

Well over 200 viral strains evolved to cause colds, most of them rhinoviruses. Only 7 strains of coronavirus infected people. 'Corona' referred to a protein complex on the surface of the virus which was thought to resemble a regal crown.

Humans first encountered coronaviruses from eating contagious cattle, roughly 3 centuries before the fall. Later outbreaks of new coronavirus strains occurred from eating infected animals.

China had an outbreak of a viral zoonosis called SARS in the winter of 98 BF. SARS was an acronym for "severe acute respiratory syndrome." CoV1, the 6th cold coronavirus, caused SARS.

SARS had high mortality rate: around 11%. But CoV1 was not very contagious. By the summer of 96 BF, SARS had run its course.

It was a brief trial run. The CoV virus was not done.

The cold called *covid* was first discovered at the end of 81 BF at a meat market in Hunan, China. CoV2, the mutant daughter of CoV1, caused covid. The animal that introduced V2 to humans was never established.

V2 was a saltation from its predecessor. For starters, V2 was far more infectious and contagious. This partly owed to ramping viral load and expelling contagious bits before causing symptoms in its victims (asymptomatic transmission). People who weren't sick were unsuspectingly spreading V2. Covid was also much milder than SARS. That helped its spread.

In its inevitable pandemic, V2 had bided its time. Thanks to international air travel, the virus had spread throughout the world months before its coming-out party in Hunan.

A doctor at the Wuhan hospital hit hardest in the initial outbreak said he and colleagues suspected the virus was highly transmissible within a month of its public emergence, weeks before Chinese authorities admitted it. That bureaucratic inertia meant that covid got a good head start in China.

Once in gear, the government imposed severe, extended lockdowns to try to stop covid contagion.

China was unable to staunch V2. But, through disciplined measures, China bettered other nations in tamping down viral spread.

Every other nation in the world got months of advance warning that their own covid epidemic was coming. Rather than prepare, governments diddled.

Then the real folly kicked in. Rich Western nations panicked when covid hit them.

Europe was supposedly a functioning supernational government: the European Union (EU). Yet, filled with dread over an unstoppable cold virus, EU coordination was utterly absent. Each nation went its own way: most to ill effect.

The senseless response to the covid pandemic was to shut down the global economy. Governments crushed their nation's vitality with mobility and social restrictions. Yet scant protection was provided for those most vulnerable to covid: in care homes and those imprisoned by the state. Further, governments failed to provide decent advice and supplies for at-home treatment.

The lockdowns and other measures that China imposed were impressive expressions of totalitarianism. If such determination had been applied to eliminating the sources of industrial self-extinction, humanity would have survived.

Covid restrictions emanated from chronic government ineptitude. Even in the richest nations, healthcare systems were fragile. In the first waves of covid, hospitals were overwhelmed. Some governments confessed that restrictions were imposed in the hope of averting healthcare system collapse.

The prudent response to the covid pandemic would have been to prepare, protect the vulnerable, and otherwise let the virus run its course. That strategy was never considered. Modern men narcissistically prided themselves with the delusion that they could control Nature.

V2 transmission was almost entirely airborne. This route of contagion was clear early on. But it took virologists months to catch on. The sophistic belief that V2 spread by contact resulted in widespread compulsive surface cleaning, to no effect. Meanwhile, the virus merrily went about its business.

V2's mildness was apparent shortly after its appearance. V2 had scant effect on those in good health. Healthy children were unaffected. At worst, V2 caused a modest cold.

By contrast, those who had not kept themselves healthy might be subject to severe covid. This was a vast majority of adults and far too many children.

At that time, 4 of 5 adults in affected nations were grossly overweight: 2 in 5 obese. In the first 2 years, covid killed 5 million.

It was not the virus that proved fatal. Instead, errant immune systems turned on their own bodies. Terminal covid was an auto-immune disease, caused by a negligent lifestyle.

Mask-wearing was mandated. Vaccines were quickly developed. Nothing thwarted V2. Instead, the wily virus swiftly guided its evolution to become even more contagious and milder. Mass vaccination only raised the bar for V2 to overcome, which it quickly did.

Deglobalization began with the inept response to the covid pandemic. Governments throughout the world pared their economies down to essentials for a year. Unemployment and attendant poverty soared. Lockdowns and restrictions in 80–79 BF shuttered global commerce and migrations.

Restrictions lasted less than 2 years. Then they were abandoned as futile. The panic subsided into resignation.

The pandemic wound down in the spring of 78 BF, after most everyone had encountered V2. The virus remained endemic, with occasional waves of new variants of V2.

V2 culled some of the unhealthiest from the global population. 5 years after the covid pandemic hit, per capita healthcare expenditures in many nations were lower than what they had been before covid struck.

Governmental reaction to the covid pandemic deepened festering societal divisions. Libertarians protested the restrictions which illiberals imposed. Citizens generally became more cynical of their government, as commanders chronically misinformed, and whatever was done was ineffectual.

The covid pandemic showed authorities as asinine in their floundering to manage Nature's dynamics. The inept response by governments to the covid pandemic heralded self-extinction.

The economic scars of covid social restrictions were considerable. The just-in-time global supply chain which had been developed over decades showed its inherent fragility. Unseen in the rich world for 4 decades, politically pausing globalization set the world on a path of chronic inflation, with effects which rippled on until the end of world civilization.

Authorities seemed to learn little from the covid pandemic. Foremost, healthcare systems were not improved. Many left the health profession after the stressful experience of governmental negligence and abuse by patients during the first years of the covid pandemic.

Covid was exemplary of a larger trend in the last century. Habitat disruptions from warming and changes in precipitation patterns aggravated the incidence of many zoonoses. Weather events brought people and morbific organisms together. For instance, floods raised the risk of encountering water-borne pathogens. Parasites enjoyed Earthers extinguishing themselves.

❀❀❀

Earther's demise reflected political will. The horrific wars of the 2nd century BF were incited by elected governments. The unrelenting drive to self-extinction was egged on by the wealthiest democracies. One need only glance at a political map of the modern world – splintered into hundreds of nations – to realize that the disciplined cooperation necessary for survival would elude Earthers.

The gyre that ended humanity was an incremental procession. The roads upon which horse-driven wagons had traveled for millennia became highways for motor cars which spewed their damaging exhausts into the skies. Nonexistent was the modern commander who had the vision to stop adoption of a technology which provided convenience but doomed future generations.

Theft from the next generation was government policy. Accumulating sovereign debt via deficit state spending was universal. Especially in democracies, politicians stayed in power by doling out money that wasn't theirs to spend.

That is not to say that the end times could not have been put off. Egregious pollution was allowed despite being recognized for what it was.

It was abundantly apparent by the early 2nd century BF that humanity was on an unsustainable path. The courage it would have took to avert the inevitable did not appear.

The reckless Roaring Twenties in the USA came to a crashing halt in 171 BF. The Great Depression ensued.

That prolonged depression came from unbridled capitalism. Its malaise lingered through the 160s BF. No systemic cure was attempted. Instead, the world's rulers fumbled for a decade before embarking on another destructive folly.

A *tipping point* is a critical moment in a complex gyre which produces an irrevocable change. For men, that year was 160 BF, when the commanders of wealthy nations engulfed the planet in a 2nd World War. Global air surface temperature shot up in 160 BF as factories geared up for mass destruction. From then there was no turning back from trends that propelled the world of men to self-immolation.

Instead of making the 2nd World War the "war to end all wars," men more tightly hewed to militarism. Extensive nuclear bomb tests explosively showed contempt for peace. Military budgets only declined for a few years after the 2nd Great War before rising again.

The continuing trend of militarism did not abate. Instead, conflicts in the 1st century BF accelerated man's demise. Men went out awash in war: combat that changed only in severity since fisticuffs were first considered a way to settle disputes.

In their own ways, road building, industrialization, urbanization, and biocides expressed unrelenting war on Nature. Even those supporting this endeavor noticed that it could not be sustained. By 141 BF, the American oil industry was aware that the unrestrained burning of fossil fuels was a global threat.

In 135 BF, an advisory committee informed USA ruler Lyndon Johnson: "Pollutants have altered on a global scale the content of the air." Summarizing the findings, the head of the American oil industry research institute foretold, "Time is running out." Nothing was done.

Formed in 132 BF, the Club of Rome was an informal league of scientists, economists, businessmen, and political leaders who pondered pressing global issues. Its focus sharpened the next year.

In 131 BF, UN ruler U Thant worried that humanity was running out of time to address its chronic problems. Nations "have perhaps 10 years left in which to subordinate their ancient quarrels and launch a global partnership to curb the arms race, to improve the *human* environment, to defuse the population explosion, and to supply the required momentum to development efforts. If not, these problems will have reached such staggering proportions that they will be beyond our capacity to control."

On the heels of U Thant's lament, in 128 BF, the Club of Rome issued its 1st report, titled *The Limits to Growth*. The authors optimistically concluded: "It is possible to alter growth trends and establish a condition of ecological and economic stability that is sustainable far into the future." Yet this "brain trust" gave no roadmap for transition.

The Club's report was in the realm of *environmental economics*. Though the desecration of Nature had been a concern to astute observers since antiquity, environmental economics only secured an academic niche in the 130s BF. Of principal concern to environmental economists were the consequences from failures in the market system.

The UN burnished the illusory glow of "sustainability" in a report published in 113 BF: "Humanity has the ability to make development sustainable to ensure that it meets the needs of the present without compromising the ability of future generations to meet their own needs."

The business of capitalism always was exploitation. As such, sustainability was framed in how much extraction and damage Nature could take before men passed the tipping point into irreversible self-extinction. Sustainability was never more than academic speculation.

Sustainability was a shifting sand. Population level, resource extraction rate, environmental degradation, and biospheric dynamics (e.g., global warming) all factored in.

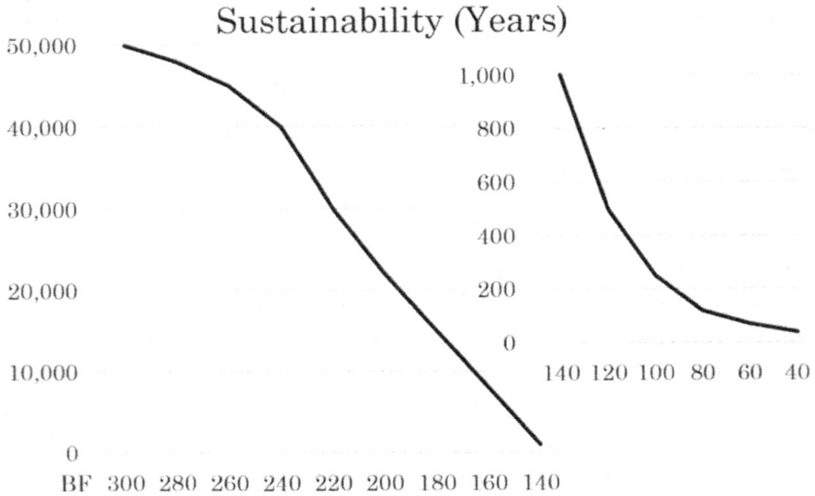


The sustainability graph dramatically highlights how quickly Earthers extinguished themselves, with no pause in their drive to self-extinction. Their fall was precipitous.

Not accounted for in the graph is the societal breakdown which transpired in the last few decades, as world civilization crumbled. Fittingly, the death of comity was the final nail in the coffin of humanity.

¤✧¤

Fossil fuel companies understood their contribution to self-extinction. In 109 BF, Shell Oil released a publicity film which acknowledged the "possibility of change too fast for life to adapt without severe dislocation."

But the rulers of fossil fuels quickly realized that such a dire message would result in an unprofitable dislocation for themselves. Their mission was to make money, not save the planet. So, the oil industry dug in their heels with denial. In 103 BF, Mobil Oil advertised: "Let's face it: the science of climate change is too uncertain to mandate a plan of action that could plunge economies into turmoil." That corporate stance stuck in plutocratic states, even after the United Nations warned that time was running out.

A summit held in Paris 85 BF brought forth pledges from 195 nations to curb emissions. Those promises were meager yet went unmet. Repeated international climate summits in the decades that followed resulted in more vows only met in the breach.

In the 1st century BF, China was the largest producer and consumer of coal, which it used to generate electricity. In 80 BF, China promised to reduce its coal habit. But expediency won out. China's economy slowed owing to its response to the covid pandemic. In 78 BF, Chinese leader Xi Jinping said that sustainability should not come at the expense of "normal life."

This sentiment – fidelity to expediency – echoed worldwide. States repeatedly backpedaled from their pledges to abandon fossil fuels. Contemporaneous with China's coal decision, the USA ramped its petrol production to counter a tiny rise in the price of oil.

The European Union had vowed to lessen Europe's warming emissions. On that they would not backtrack. But there was no way that Europe was going to meet that promise. So, in 78 BF, the EU reclassified methane, a potent greenhouse gas, as environmentally benign ("sustainable"). EU ruler Ursula von der Leyen stated the priority: to "protect our single market and industrial supply chains from disruptions." Such expediency made a mockery of addressing global warming.

Public alarm over self-extinction did not erupt until the mid-70s BF, when it was already far too late. This owed to studied ignorance.

By the late 2nd century BF, the continuing cruelty of an exploitative political economy and indifference by authorities to injustice, poverty, and suffering made reading the news a dreary experience. The situation only worsened in the 1st century BF, as reports of civil discord, pollution, and catastrophe became more frequent and disheartening.

Many gave up on reading newspapers. "People need to have something to believe in. People need to have a sense of possibility," said 1st-century-BF journalist David Bornstein. But the function of news was to inform, not appease. Bornstein's sentiment was symptomatic of the lack of fortitude common among peoples in post-industrial nations.

There was a widespread sense of entitlement among the middle and upper classes. Though aware of the massive desecration of Nature, very few changed their habits of unsustainable consumption. "Enjoy it while it lasts" was the tacit logic. Meanwhile, those in the lower class struggled as they had throughout history.

Therein lay the core problem of democracy, which relied upon an informed electorate as a platform for choosing apt leaders. In the last century, as the global situation grew more dire, political regimes stayed conservative because electorates preferred the false promises of politicians pledging to bring back "the good old days."

Despite the extensive litany of destruction, the unfolding mass extinction event was invariably labeled "climate change." Extreme weather events around the world increased in frequency and intensity: heat waves, wildfires, flash floods, drought, dying crops.

Yet global media still bubbled with cautious optimism. There was time for remedy. Incremental countermeasures might suffice. So-called "green" technologies were fawned upon, especially solar panels and electric cars.

A bold few suggested systemic change to the global political economy. Such prophetic radicalism went nowhere.

The death toll ratcheted into a higher gear in the 3rd decade of the last century. Even then, most people simply did not think self-extinction possible. Those concerned with the possibility still considered it a distant prospect.

The only real hope of averting self-extinction would have been to recognize the nature and scope of the dilemma and systemically address it. Instead, people invested in ignorance and fantasy.

Authorities hid their nation's polluting emissions. They did so by tacit agreement among the most polluting countries, notably the USA, China, India, and Australia. As the methane example illustrates, Europe joined in.

Pledges to do something substantive about global warming were based upon falsified data. This deception was noted by UN head António Guterres in 78 BF: "Government and business leaders are saying one thing but doing another. Simply put, they are lying. And the results will be catastrophic."

Greenhouse Gas Emissions & Global Warming

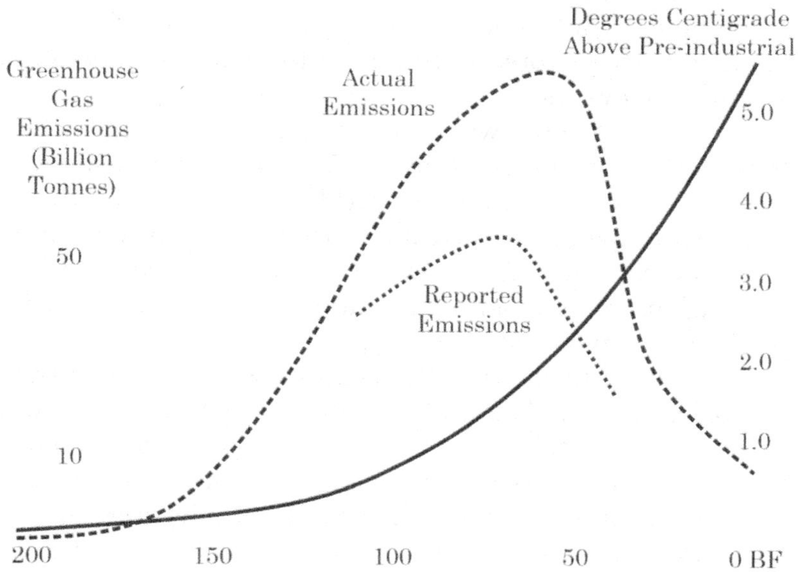


The gap between reported and actual emissions grew alongside public alarm. By 75 BF, reporting of warming emissions was at least 1/3rd less than actual.

The successful campaign to species suicide was spearheaded by conservative politicians, abetted by corporate media. They deceived and a gullible public believed. The result was inaction until it was far too late to effect a cure.

The corporate press was beholden to the status quo by dint of being owned by those who profited from keeping things as they were. They used trickery to mislead their readers: reporting events yet understating the apparent trend. Even into the late 70s BF, corporate media stuck to the theme that there was hope for a better future. This concerted deception made the collapse of civilization seem sudden to hapless humans comforted by their conformity.

From the late 2nd century BF, climatologists employed mathematical models to track the pace of global warming. They consistently projected less warming than what occurred. This serial conservatism owed partly to relying upon rigged data, partly to the models, and partly to the modelers themselves.

As computers grew more powerful, climate models evolved to greater complexity. This afforded a better handle on dynamics. But models still underestimated feedbacks among climate factors, and thereby the pace of change and harshness of events.

For instance, what was going on in the upper atmosphere affected the rate of polar melt as well as the severity of storms, floods, and droughts worldwide. But modeling the stratosphere took cracking computations that taxed even the most powerful machines. As such, climate models understated what was happening at the poles and the increasing toll of destructive weather events.

In the early 1st century BF, polar ice was melting at twice the rate predicted. The underestimate owed to the perspective employed. Satellites in space were being used to monitor the extent of changes at the poles. This aerial view missed massive melt from underneath, relentlessly chewing away at glacial foundations. As such, climate models did not sufficiently factor in basal thaw, including the feedbacks inherent in that gnawing dynamic.

Climate models were the crystal balls which affected public policy: especially from the 70s BF on, as the economic knocks of global warming became more pronounced. Politicians had enough to deal with without predictions of impending demise breathing down their necks. To those in power, doomsaying was damnable.

Modelers considered a wide range of scenarios. Whereas models' worst-case projections more closely corresponded to what happened, modelers chose more moderate predictions as likely. Climate modelers seemed loath to upset their paymasters with alarming forecasts. John Fyfe was typical of climatologists when he cautioned in 80 BF that "it's a bit too early to get wound up" about global warming. Being blasé was jolted away 2 years later.

78 BF was a watershed year in extreme weather events. Heat waves, wildfires, storms, floods, and droughts hit hard around the world. Glacial ice was melting at a startling pace. This prompted climatologist Joerg Schaefer to say: "All of the predictions are way too conservative. The change will be much faster."

The pollution and waste which propelled self-extinction were inherent to plutocratic capitalism. Rulers knew that they were murdering humanity but did it anyway. Even when tens of millions were dying annually from the consequences, the drive toward demise stayed stuck in high gear.

"Capitalism is the extraordinary belief that the nastiest of men for the nastiest of motives will somehow work for the benefit of all," marveled English economist John Maynard Keynes in the early 2nd century BF. Mankind propelled by myopic selfishness could not possibly have a happy ending.

The pressing global dilemma necessitated a coordinated reversal in the political economy which was driving humanity to a dead end. What happened instead was ineffectual.

From the 2nd decade of the last century, politicians felt rising pressure to address "climate change." Much lip service was paid but no remedy proffered. The covid pandemic had already shown with how little wisdom the world was governed. Plutocracy retained its grip.

The sensible response to save humanity was sustainable socialism. That turnabout was untenable.

The time horizon for politicians in democracies was whatever the local election cycle was. Staying in power was paramount. That meant expediency always trumped solution.

Nationalization and dictatorial rule were tried several times in the modern epoch. The only autocrat to artfully pull it off was Josip Tito of Yugoslavia, in the mid-2nd century BF. Tito's objective in crafting a command economy was to make Yugoslavia prosper. He did so through increased industrialization. Tito was no role model for how man might overcome its chronic overexploitation of Nature.

By contrast, Cuba exemplified a country wobbling in the right direction. Fidel Castro took control of Cuba in 141 BF. He dictated for the next half-century, during which the country economically and socially struggled. The USA, a natural trading partner, kept an embargo against Cuba because of Castro's socialism.

Castro embraced environmentalism, though he never managed to forge an economy which was ecologically sustainable. Continuing political dissent suggests that Cuban society never accepted Castro's vision. Castro's heavy-handed governing style did not help his cause. After Castro died, Cuba inched back toward a market system as the nation economically and socially struggled.

In the last century, Xi Jinping of China could not have been better poised to shift that state toward survival. He demurred from the Herculean task, confining himself to measures which entrenched his power and expanded China's geopolitical influence.

A few nations making the 'right' move was never going to be enough. To "save the planet" would have required a global effort of economic restraint, along with systematic and systemic coordination. Such an endeavor was not beyond man's capacity, but it was well outside their willingness to try.

Voters in democracies continued to elect conservative governments through the 70s BF. Rather than sensibly lead the way, politicians managed their constituents' myopia by appealing to sentiments which kept the status quo entrenched. Exploitative capitalism continued its dominion.

The very idea of rejecting capitalism for state-run socialism was beyond the pale for majorities in electorates, even those who would have greatly benefited from it.

Conservatism was a stolid political bedrock of inertia. Its foundation was a stiff mixture of selfishness, fear, gullibility, hopeful ignorance, and belief in plutocracy.

The USA was exemplary in its deep distaste for equitable societal management. The dominant tribe of ill-educated whites feared usurpation of their station from those they had long oppressed. The dominance of American conservatism only broke in a wave of panic toward the middle of the last century.

From the early 60s BF, around the world, the bitter fruits of capitalism were met with growing public clamor for radical action. This global revulsion owed to youth whose futures had been robbed, and a mounting death toll, which was over a hundred million people per year by the close of that decade.

Strident protests against the system for self-extinction were first met with brutality by police. That changed in the 50s BF, as minions of the state realized their predicament: trying to protect the power of an unsustainable status quo. The breakdown in civil order spelt collapse.

❀❀❀

Humans lived during an era of low sea levels, unmatched for 190 million years. Global sea level rise noticeably began in 237 BF, just as industrialization was starting. The lift first owed to warming water. The rise was then augmented by glacial melt.

Marine expansion was inexorable. Soaring global warming in the last century accelerated it.

Rising seas increasingly flooded ocean islands and continental coastal areas. East Asia and the southern Indian subcontinent were the worst affected.

Founded in 32nd century BF, the city of Venice, Italy, was built upon 118 small islands in a shallow lagoon. The islands were separated by canals and linked by over 400 bridges.

During the Middle Ages, wealthy Venice was an intercontinental trade hub, the 1st international financial center, and a locus of exquisite art and music. Venice was the most beautiful city Earthers ever built. Beset by worsening floods in the last 2 centuries, submerging Venice was abandoned in the mid-1st century BF.

Dozens of coastal and island cities sank faster than the rate of sea level rise. That owed to groundwater draw dropping elevation.

Jakarta, Indonesia, was home to 10 million people in 80 BF. By that time, its air quality was among the worst of national capitals in the world. Considering how unbreathable air had become in many places, this was one mean feat. But bad air was not what did Jakarta in.

Jakarta was sinking 25 centimeters per year as its people unsustainably drew water from underground aquifers. This, combined with rising sea levels, sank Jakarta.

The Indonesian government did not try to save Jakarta. Instead, in 74 BF, the capital was moved to a newly built city on another island over 1,000 kilometers away. Like Venice, Jakarta was abandoned by the mid-1st century BF.

Island cities were not alone in being inundated. Coastal cities in China and India were done in by storm-induced flooding.

Warmer air holds more water vapor than cooler air. Wind-driven storms aside, rainfalls became increasingly intense, flooding low-lying areas. A perennially popular locale for settlements, flood plains earned their name in the last century. Water saturation was compounded by an urban heat-island effect: concrete-laden cities warm the air around them, urging fierce rainstorms.

Numerous cities in the south China coastal province of Guangdong were lost to flooding. Guangdong had been a maritime trading center since antiquity. By the late 2nd century BF, Guangdong had also become the most populous province in China.

Mumbai was a city on the west coast of India, facing the Arabian Sea. Mumbai began in the 25th century BF. In 255 BF, this archipelago of 7 islands artificially became a single landmass via landfill.

By the last century, Mumbai was the 2nd-most populous city in India. Thanks to unchecked pollution, Mumbai's air was barely breathable.

The Mumbai monsoon season ran between June and September. The city often flooded then from the 2nd century BF on. Tropical storms repeatedly pummeled Mumbai with increasing intensity in the last century. Flooding became chronic. Mumbai was abandoned mid-1st century BF.

On the opposite side of India from Mumbai was the Bengal biome: a flood plain facing the Bay of Bengal. Fed by the Ganges and other rivers, this lush, tropical region was the largest river delta in the world. Bengal comprised coastal east India (West Bengal) and Bangladesh.

Bengal had been populated for hundreds of thousands of years. By the 1st century BF, people were crammed into dense seaport cities.

Storms and flooding worsened in Bengal in the last century. As a result, some 250 million people there were displaced or died prematurely by the end of the 30s BF.

Cities on the east and southern coasts of the USA were increasingly battered by storms and attendant flooding. Among others, the southern cities of New Orleans and Miami were abandoned by mid-1st century BF. The economic collapse of New York City in the late 40s BF owed partly to storm damage and flooding.

In the last 8 decades of Earther existence, island nations throughout Oceania were repeatedly pummeled by increasingly severe storms. Once episodic, flooding became chronic.

Hundreds of Polynesian islands with sizable populations were cleansed of human habitation by 65 BF. A modest rise in sea level by that time brought "king" tides which deposited salt water inland, making groundwater undrinkable and salting vegetation to death. Seasonal cyclones of increasing intensity further made such islands uninhabitable. Many Polynesians migrated before then.

Sea levels rose for hundreds of millennia after the fall. The oceans swallowed thousands of islands and overran low-lying continental shelves. Sea levels peaked at a height last seen 50 million years earlier. Along with climate, high sea level affected biomes worldwide, and subsequently the evolution of the animals and plants which survived the Earther mass extinction event.

¤✧¤

Heating Earth's oceans put pressure on its nether regions. The upshot was tremorous and volcanic. "Water largely initiates and fuels volcanic eruptions," explained 1st-century-BF volcanologist Dan Rasmussen.

Eruptions and earthquakes became more frequent in the last century, and even worse in the millennia that followed Earther's demise.

Geophysical violence ruined the habitability of numerous Pacific islands. Japan was one of them.

Japan spanned an archipelago of 6,852 islands, with 5 main islands. It was densely urbanized.

Japan sat on or near the boundaries of 4 tectonic plates. The most active was the Nankai Trough, a subduction zone just south of the Tokyo region. In the trough was the Kumano

Pluton: a huge boulder that amplified the turbulence of marine heating.

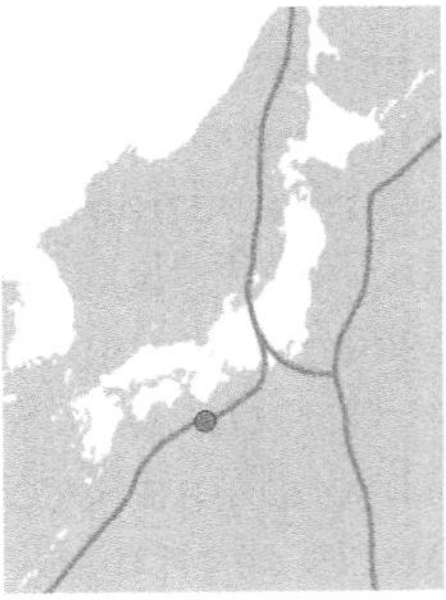

Unsurprisingly, Japan had a long history of earthquakes and huge waves known as tsunamis. *Tsunami* is Japanese for "harbor wave."

Japan was wracked by quakes and tsunamis before world civilization collapsed. The resultant infrastructure destruction accelerated the demise of that country.

¤✧¤

In the finale of humanity, Earth's biomes turned adverse. Bursts of extreme heat intensified. Polar ice profusely melted. But the hotting up was the worst in the tropics. Drought, wildfires, and severe storms added to the thermal toll.

Dust alters biomes. Dust affects cloud formation. Dust from north Africa and Arabia intensified monsoons in India. For thousands of years, iron-laden dust from the Sahara Desert in north Africa fertilized the Amazon basin and the nutrient-poor wetlands of south Florida.

1st-century-BF desertification created vast fields of dust. Sandstorms worsened in the last century, taking many lives in north Africa, the Middle East, central Asia, and China.

Refugee migrations from wars and political repression were a historical continuum. These worsened in the last century. What supplemented the refugee waves were people fleeing inhospitable habitats, water shortages, and famine.

Indecency by authorities in coping with refugees added to the death toll and societal breakdown. Nationalism and racism (by dominant tribes) begat isolationism when it came to migrants, who were often refused or otherwise herded into the purgatory of detention camps.

Global food yields peaked in the early 1st century BF. Crop failures and outright depopulation on a mass scale became regular events from the late 70s BF. Heatwaves, drought, flooding, and storms all played their part.

As California exemplified, projects to preserve water supplies or relocate were scattershot and haphazard, even though there was advance warning for at least a half century before the problem became critical.

The human population of the Middle East – the cradle of Western civilization – peaked around 80 BF at ~412 million people. By 70 BF this region was roasting year-round. 2 decades later, by 50 BF, 70% of the population had succumbed. The rest migrated or died off within the next quarter-century.

The Indian subcontinent comprised India, Pakistan, and Bangladesh. In the century and a half before the fall, the region's people tripled. Population peaked at 1.8 billion people around 65 BF. 20 years later, the population had halved. In the last years, fewer than a million survived, in remote areas at higher elevations (the Himalayas).

The Indian subcontinent followed the thermal arc of the Middle East, albeit abetted by incredible levels of air and water pollution, especially in urban areas. Torrid heatwaves assailed the Indian subcontinent from the early 1st century BF.

In the 70s BF, huge mountains of garbage near Delhi, India's capital, spontaneously combusted during scorching heatwaves. The fires started in methane pockets from organic decomposition. These towering infernos burned for weeks, releasing huge clouds of noxious smoke. Other cities in India also had such rubbish fires.

Storms and drought also took their toll on Indian cities, as did freshwater shortages in many areas. Meanwhile, coastal areas went underwater.

India was the 3rd biggest contributor to global warming, behind the USA and China. Even toward the mid-1st century BF, India made scant reduction in coal-fired power plants and vehicle emissions. And the state persistently lied about its environmental circumstance.

India's government was indelibly corrupt. It engendered religious violence (Hindu on Muslim), maintained casteism, and spent exorbitantly on armaments, all the while leaving most of its people impoverished.

As in many other nations, India's fall was marked by riots and brutal police reaction. On its way out, India tore itself apart with savagery. It was a vicious terminus for a country that began with strong spirituality.

The USA also met a violent end. After extensive rioting over shortages, the nation collapsed into a brief civil war which more resembled serial massacres. A militaristic minority – right-wing whites – were heavily armed. They coalesced into roving gangs. The last survivors scrounged packaged foods and drank bottled water until disease or starvation took them.

Europe's end was more sedate. Though societally significant, rioting played a minor part. Governments lasted longer in Europe than in the USA. Unsurprisingly, the disciplined northern nations fared better than southern Europe, which descended into terminal chaos.

In the last 2 decades, only a few island nations managed to sustain a semblance of civility. New Zealand was one of the last places where Earthers lived that way.

Neighboring Australia, chronically mismanaged and ecologically abused, succumbed earlier. Its last settlements were in coastal areas at higher elevations, and in Tasmania, the large island south of the main continent. As with much of the still-habitable world, crop failures were instrumental in Australia's demise.

¤ ✧ ¤

Earther's end was death by a thousand self-imposed cuts. What made their extinction inevitable was institutional inertia: hewing to an unsustainable political economy. Plutocracy ruled until money lost its power. Humanity choked on its own conservatism.

Humanity's last, best hope to overcome its terminal malaise was education: the key opportunity which altriciality afforded. Instead, Collective wisdom dimmed. Western societies exemplified loss of acumen. Rationality had played a larger role in public discourse during industrialization, from the mid-3rd century BF. Use of words associated with logic rose, such as "determine" and "conclusion."

This rise in reasoning reversed from the 110s BF, when emotive statements, such as "feel" and "believe," revived as parlance. "This change accelerated around 2007 [93 BF], when the frequency of fact-related words dropped, while emotion-laden language surged: a trend paralleled by a shift from collectivistic to individualistic language," observed 1st-century-BF sociologist Marten Scheffer.

The civilization of men was a thin veneer. Despite the guise of democracy, nations held together not by consensus, but by the state's implicit threat of violence against those who contravened its command. When disorder became the norm, the power of the state evaporated.

The collapse of institutions worldwide resulted in rapid decimation. The spasm of violence that attended governments losing their grip cost over a billion lives. This was a sordid refraction of how civilizations had first formed: subjugation under the threat of death.

Scattered pockets of people who lived in high latitudes survived the demise of civilization for a few decades. Many of them were autochthones. They had lived close to Nature and were fortunate to be in a habitat that still offered clean water and a food supply, including the ability to grow crops. Many of these survivors lived in remote mountainous regions.

From a geological perspective, Earther's manufactured mass extinction event was sharp and short. Though there was a momentum of unsustainability from antiquity, men wiped themselves out with their industrial advance in a mere 3 centuries.

Earthers drove many other animals extinct as they were doing themselves in. By the fall, many land animal populations were down 99% from what they had been only 3 centuries earlier, before industrialization accelerated the demise. The marine death toll was even worse.

Earther's engine of pollution died when the machines fell silent. Lowered albedo and increased volcanic activity extended the warming effect. But, within a half million years, tropical biomes were again becoming tolerable.

As with other mass extinction events on Earth, plants were not as severely affected as animals, and their recovery quicker. The return of vibrant land animal life took much longer to get into swing: nearly 1 million years. Considering how severe the loss of life had been, recuperation was remarkably swift.

¤✧¤

The passing of man ended a planetary plague. Hominids were survived by their cleverer cousins: rodents, the first mammals. In the pecking order of animals, Earth reverted to a semblance of its situation before dinosaur demise.

Only 4 groups of terrestrial vertebrates evolved on Earth: amphibians, reptiles, mammals, and birds. The outlier was the one with feathers and wings, and a brain unlike any other. The birds of Earth are a highly distinctive animal.

Avians were an adept adaptation from larger dinosaurs. They emerged in their refined form 160 MBF, following 50 million years of experimental evolution.

Birds became itty-bitty by skipping the juvenile growth spurt that gave dinosaurs their girth. Like hominids, birds descended via neoteny.

In size, avians appeared a reversion back to the earliest dinosaurs, who were quite small. Miniaturization afforded flight and the novel opportunities that went with it.

Birds were sorely hit by the Earther mass extinction event. Destruction of their habitats and food supplies downed birds to scant species in small numbers. But they endured.

The rejuvenation of birds was a phoenixity. Avian adaptability was their redemption.

The radiation of birds in the post-Earther era was an echo of post-dinosaur radiation, 65 million years earlier. The avian species that survived recovered and evolved. There was a flurry of new birds and mammals to take advantage of the hospitable habitats that plants were providing.

The absence of men allowed the widespread revival of robust flightless birds. The secret of their success owed in large part to what men lacked: restraint, and a conviviality that let them live peaceably together in grand flocks.

# Glossary

~ : approximately.

## #

*3-age system*: a tripartite archeological sequential periodization of human prehistory and early history, comprising the Stone Age, Bronze Age, and Iron Age. Never mind the Wood Age.

*3rd party*: an individual or group not directly involved in an issue (typically a contention). The 1st party is the initiator. The 2nd party is the object of the 1st party.

## A

*actuality*: the world experienced through the senses. Compare *reality*.

*adaptation* (evolutionary biology): the teleological process of adjusting to ecological circumstance.

*Age of Discovery* (aka *Age of Imperialism*) (7th–4th century BF): the period that Europeans came to comprehend the vastness of Earth through naval exploration, and to conquer and colonize foreign lands.

*agriculture*: the cultivation of one life form by another, most commonly referring to growing plants.

*albedo*: the ratio of light reflected by a surface to that received by it.

*altricial*: animals that are relatively immature and immobile at birth or hatching and so require parental care. Many mammals are altricial. Contrast *precocial*.

*Anatolia* (aka *Asia Minor*): the westernmost protrusion of Asia, comprising most of modern-day Turkey. Anatolia is bounded by the Black Sea to the north, the Mediterranean Sea to the south and the Aegean Sea to the west. The eastern border of Anatolia was a line between the Gulf of Alexandretta and the Black Sea, east of the Armenian highlands.

*angiosperm*: a plant that seasonally exhibits flowers. Angiosperms arose 245 MBF, incorporating several innovations, including leaves, pollen, flowers, and fruit. A wave of angiosperm radiative speciation began 144 MBF.

*animal*: a kingdom of eukaryotic heterotrophs. Most animals are motile. Compare *plant*.

*animism*: the doctrine that that there is no separation between the physical and spiritual world, and that a vital energetic force is inherent in all of Nature. Compare *vitalism*.

*anthropoid*: an apelike primate. Compare *hominid, humanoid*.

*anthropology*: the study of humans.

*ape* (aka *great ape*): a tailless primate; not a monkey.

*Asia*: a large landmass continent, bounded by Europe and the Arctic, Pacific, and Indian oceans.

*atmosphere*: the layer of gases that surround a body of sufficient mass. The atmosphere is held in place by the gravity of the body.

*auroch*: the ancestor of modern domestic cattle.

*Australasia*: a region of Oceania comprising New Guinea, Australia, New Zealand, and neighboring islands.

*Australopithecus* (4.2–1.8 MBF): a relatively long-lived genus of largely vegan hominids with considerable species diversity.

*authority* (political): an individual or group with the power to force submission from others.

*autochthone* (aka *aborigine, native*): an original inhabitant of a region.

*autocracy*: a state commanded by an autocrat. Contrast *democracy*.

*autocrat* (aka *dictator, despot*): a commander with unlimited authority. Compare *ruler*, *monarch*. See *commander*.

*autopoiesis*: a dynamic of self-sustaining activity; a system capable of maintaining and reproducing itself. A biological cell sustaining itself is an example of autopoiesis. Compare *homeostasis*.

*autotroph*: an organism that makes its own food. Contrast *heterotroph*.

*Axial Age* (30th–25th century BF): 2nd-century-BF philosopher Karl Jaspers' proposal that there was a pivotal age in world history regarding philosophy and religion.

## B

*baboon*: a genus (*Papio*) of monkeys with strict social dominance hierarchies.

*bad* (economics): an externality from a good. See *good*.

*belief*: a habit of the mind to axiomatically treat ideas as true; confidence in abstractions as real.

*BF*: before the fall. A reverse chronology, beginning with 0 (zero) from Earther Gregorian calendar year 2100 (the beginning of the 22nd century BF) and running backward in time.

*biocide*: a chemical agent to kill animals (pesticide), plants (herbicide), or fungi (fungicide).

*bioelement*: a planetary ecological element. The bioelements include the *atmosphere*, *lithosphere*, *hydrosphere*, and *biota*.

*biome:* an area where organisms live with similar conditions, both geographically and climatically.

*biosphere*: the global summation of the Earth's ecosystems.

*biota*: the life forms (i.e., organisms and viruses) in an environment.

*bird*: a feathered, bipedal, egg-laying vertebrate. Birds descended from dinosaurs.

*black* (sociology): modern vernacular for dark-skinned people; especially in the USA & UK, those with an African heritage. Owing to a legacy of slavery, blacks were subordinates in modern societies where whites were dominant. Contrast *white*.

*bolide*: a meteorite; a brighter-than-usual meteor; officially defined from a perspective on Earth as a fireball brighter than any of the planets.

*Bronze Age* (~5400–3400 BF): the middle period of the 3-age system, noted for the metallurgical production of bronze; the *Stone Age* preceded, the *Iron Age* followed.

*Byzantine Empire* (1770–647 BF): the predominantly Greek-speaking continuation of the eastern Roman Empire until annexed by the Ottomans in 647 BF.

## C

*capital*: financial wealth that can be employed to produce more wealth.

*capitalism*: a market-based economic system based upon private ownership of resources and their exploitation for exclusive profit. Compare *socialism*.

*carbon* (*C*): the element with atomic number 6; an extremely friendly element, with 4 electrons available to form covalent bonds. Life is based upon molecules made with a carbon backbone.

*carbon cycle*: the gaseous cycling of carbon exchange among the geosphere (deep Earth), pedosphere (soil), hydrosphere (water bodies), atmosphere, and biosphere (living ecological systems).

*Cenozoic* (65.5–0 MBF): the geological era from the demise of the dinosaurs to the self-extinction of hominids.

*chordate* (Chordata phylum): an animal with tadpole characteristics at some developmental stage. Chordates include lancets, tunicates, fish, amphibians, reptiles, birds, and mammals.

*clade* (biological classification): a group of biological taxa, such as genus, which includes all descendants of a common ancestor.

*clan*: a group of families or people of common descent. Compare *tribe*.

*climate*: a characterization of tropospheric activity in an area over 3 decades, accounting for seasonal variations. The standard of 30 years was often adjusted to suit reportage. Compare *weather*.

*clinker*: a kilned then quenched cement product based upon limestone; used to make *Portland cement*, the common modern cement.

*coal*: a blackish combustible sedimentary rock from naturally compressed and heated vegetation. Compare *petroleum*.

*Ĉocö*: the universal, unified field of Cönsciousness/cöherence which localizes into individualized consciousnesses, minds, and the perception of Nature.

*cöherence*: the intelligent interaction behind Nature. Like Cönsciousness, cöherence localizes.

*Collective*: the 99.9+% of modern Earthers who followed their biological urges as natural imperative. The Collective were slaves to their minds, and thereby iğnorant. The Collective cherished emotions and beliefs. As believers in dualism or matterism and in taking existence at face value, the Collective were naïve realists.

*command economy*: a state-run, organized production and distribution system. Compare *market economy*, *traditional economy*.

*commander*: a controlling authority of a state. See *ruler, autocrat, monarch.*

*commune* (politics): a European medieval city-state political regime of oligarchy.

*consciousness*: the platform for awareness in an individual life constituent, such as a protein, virus, cell, or organism. Compare *Cönsciousness.*

*Cönsciousness*: the unified field of consciousness. Cönsciousness naturally localizes into individualized consciousnesses. Compare *consciousness.*

*conspecific*: of the same species. Contrast *interspecific.*

*construal*: an interpretation.

*coremind*: fundamental perception. Compare *nattermind, willmind.*

*corporation*: an economic group legally authorized to act as an individual.

*corvid*: a cosmopolitan bird family (the crow family) which had over 120 species during the era of humans.

*coronavirus*: a virus which may cause a "cold" (upper respiratory tract) disease in humans, so named because of a crown-like protein structure on the surface of the virus. See *covid.*

*corporatism*: sociopolitical organization of a society by major corporate interest groups.

*covid*: the "cold" disease caused by the coronavirus CoV2.

*cultivar*: a variety of a cultivated plant.

*culture*: the shared language, customs, beliefs, cuisine, and art of a social group.

## D

*Dark Ages*: the 17th–12th centuries BF in Europe; the early Middle Ages, following the decline of the Roman Empire. The term *Dark Ages* was generally disparaged by modern historians, owing to its critical overtone. Yet the aptness of its cultural attribution cannot be denied. Italian scholar Francesco Petrarch coined the metaphor in the 770s BF, when writing of the previous historical period: "Amid the errors there shone forth men of genius; no less keen were their eyes, although they were surrounded by darkness and dense gloom."

*democracy*: a political system where an electorate elected rulers. An electorate was a select group of voters. The viability of democracy rested upon the assumption that the average voter was sensible enough to select societal commanders. Self-extinction showed that democracy was a failed experiment. Contrast *autocracy*.

*deontology* (philosophy): measuring morality by inherent goodness, not result. See *moral absolutism*. Contrast *teleology*.

*dictator*: an autocrat.

*dinosaur*: a diverse clade of reptiles, excepting birds – which is an arbitrary exclusion, as birds descended from dinosaurs.

*DNA* (deoxyribonucleic acid): a complex molecule supposedly employed as a physical template for biomolecular production.

*dualism*: the metaphysical belief that reality is bifurcated between the physical and the mental. Contrast *monism*.

*dynasty*: a succession of monarchs, where the monarch has the power to choose or influence their successor.

## E

*Earth*: the 3rd planet from the Sun; the densest and 5th largest.

*Earther*: a human on Earth.

*ecology*: as a subdiscipline of biology focused on patterns of interrelations between life forms (e.g., cells, organisms) and their habitats, including other organisms; more broadly, the relations between bioelements.

*economic*: the idea that objects and altruistic behaviors are goods and services respectively, as part of a materialist value system.

*economic system*: an overarching characterization of how an economy works. There are 3 main economic systems: traditional, command, and market. A traditional economy is tribal, with scant surplus. A command economy is a planned economy, typically under centralized control. A market economy is, ostensibly, a disorganized free-for-all centered on trade. See *economy*.

*economy* (economics): a societal economic gyre.

*ecosystem*: the community of biota in a biome, and the abiotic (non-living) elements within the area.

*endogamy*: mating within one's clan. Contrast *exogamy*.

*endowment effect* (aka *divestiture aversion*): sentimental attachment to objects above their economic replacement value.

*energy* (physics): the idea of an immaterial force acting upon or producing matter. Energy is characterized relatively and by type: how it affects matter. Energy manifests only through its effect on matter. Though the foundational construct of existence, energy itself does not exist. As matter is made of energy, the very characterization of energy tidily proves *energyism*.

*energyism*: the monistic doctrine that Nature is a figment of the mind. Energyism differentiates between actuality and reality. Whereas actuality is phenomenal (perceived), reality has a noumenal substrate, emergently spawning a shared actuality (showtivity) via a unified Ĉocö. Contrast *matterism*.

*enlightenment*: the 1st elevated level of consciousness in which contentment reigns. Compare *iğnorance*. See *unity consciousness*.

*environment*: a designated spatial region or conceptual realm.

*eukaryote*: an organism with cell structures (organelles) separated by membranes. Multicellular life is eukaryotic. Compare *prokaryote*.

*Eurasia*: Earth's largest landmass in the era of apes, comprising Europe and Asia, including Borneo and other nearby islands. Compare *Australasia*.

*Europe*: a continent west of Asia, separated from Asia by the Ural Mountains, and otherwise bounded by the Atlantic and Arctic oceans and the Mediterranean Sea.

*eusocial*: an animal that has: 1) overlapping generations, 2) cooperative care of the young, and 3) reproductive division of labor. Eusocial animals tend to cooperation generally.

*externality* (economics): an unintended byproduct of an ecological activity, typically making an economic product. Waste and pollution are exemplary externalities.

*evolution* (evolutionary biology): the process of adaptation, most apparently seen as a distinctive change across successive generations of a population.

*existence*: what appears in the mind (perceived) as a material world. See *actuality*.

*exogamy*: mating outside of one's clan. Contrast *endogamy*.

## F

*fact*: recall of an experienced event.

*Fertile Crescent*: the geographic area from the upper Nile River in Egypt through the Middle East to the Persian Gulf, including the regions of Mesopotamia and the Levant.

*feudalism*: a societal system prevalent in medieval Europe, with socioeconomic hierarchy based upon land holding.

*feudal*: a social control system based on fief; regional authority.

*fluid sociality*: a dynamic social group comprising a larger community with subgroups, including families and close friendships. Groups merged for safety, such as for sleeping or in breeding colonies, but then split up for foraging during the day. Various animals have fluid sociality, including fish (guppies), cetaceans (dolphins), ungulates (deer), elephants, most mammalian carnivores (lions, hyenas), and primates.

*fossil fuel*: a flammable fuel formed from dead organisms. Coal, petroleum, and methane are fossil fuels. Coal comprises ancient dead plants pressed into rock resemblance. Petroleum originates from archaic algae and zooplankton, turned into a viscous brew. Methane is a gas. Fossil fuels take tens of millions of years to form, and so, in their extraction, are nonrenewable resources.

## G

*genus* (plural: *genera*): a category of organisms, more generic than *species*.

*glacial period* (aka *glaciation*): a period of glaciers, typically thousands of years, within an ice age, marked by colder temperatures and glacial advances. By contrast, interglacials are periods of warmer climate within an ice age. The last glacial period in the hominid era ended 15 TBF.

*good* (economics): a product or service. See *bad*.

*government*: the authorities responsible for governance of a political unit (e.g., nation).

*grass*: a large, versatile, ubiquitous monocot that grew on all of Earth's continents. Grasses had small flowers and sheathing leaves covering hollow stems. They included cereals and bamboo, but not other plants commonly called grasses, such as seagrasses, rushes, and sedges (though rushes and sedges are related to grass).

*Great Depression* (171–161 BF): the longest and most severe economic downturn in the modern world before the fall. The financial shock that sparked the depression originated in the United States. Its contagion quickly spread to Europe.

*Great Dying* (252 MBF): Earth's most severe mass extinction event prior to Earther self-extinction.

*greenhouse*: an enclosed area in which a desired temperature range is maintained, usually with much glass to let abundant sunlight in. Greenhouses were used to cultivate tender plants or grow them out of season.

*greenhouse* (climate): see *hothouse*.

*greenhouse effect*: the process by which radiation from the atmosphere warms a planet's surface.

*greenhouse gas*: an atmospheric gas that has a warming effect on a planet's surface from shedding heat energy absorbed via infrared radiation. The primary greenhouse gases in Earth's atmosphere are carbon dioxide, methane, nitrous oxide, and ozone. Water vapor acts as a greenhouse gas.

*groupthink*: the praxis of approaching and dealing with issues via consensus. Democracy is a political form of groupthink.

*gyre*: a conceptual framework which treats a physical system as a dynamic vortex. A gyre may be characterized by its structure, qualities, thermodynamics, and interactions.

## H

*habitat*: the environment in which a species population lives.

*heat engine*: a machine that transforms input thermal energy into mechanical power.

*herbivory*: feeding on plants.

*heterotroph*: an organism that cannot make its own food. All animals are heterotrophs. Compare *autotroph*.

*Himalayas*: a young mountain range formed by the Indian subcontinent moving north and slamming into Eurasia. 9 of the 10 highest peaks on Earth are in the Himalayas.

*Hinduism*: the dominant religion of India's natives, founded upon some of the earliest texts on spiritual realization, when the Indus civilization was extant.

*homeostasis* (biology): a bodily regulatory process of striving for holistic health. Compare *autopoiesis*.

*homeostasis* (physics): a tendency toward systemic stability.

*hominid*: an ape descendant, some of which became hominin.

*hominin*: the hypothesized clade that descended into humans. See *hominid*.

*Homo* (from 2.4 BF): a diverse genus of hominids which included modern humans (Earthers).

*hothouse* (aka *greenhouse*): a duration lasting millions of years where Earth is warm and typically humid, completely lacking continental glaciers. Contrast *icehouse*.

*human* (aka *Homo sapiens, Earther*): a bipedal, largely furless primate.

*humanoid*: an ape with human-like characteristics.

*hydrocarbon*: an organic compound comprising hydrogen (H) and carbon (C). Methane ($CH_4$) is an exemplary hydrocarbon.

*hydrogen (H)*: the primordial chemical element, with a single proton and solitary electron; the lightest and most abundant chemical in the universe, comprising 75% of cosmic mass.

*hydrosphere*: the bioelement of water, including the participants in the water cycle.

*hypothesis*: a guess gussied up in scientific garb. Under the scientific method, hypotheses are ripe for falsifiability testing. Compare *theory*.

## I

*icehouse* (aka *ice age)*: a span of millions of years where the world has continental ice sheets, tending toward cool and arid climate. Contrast *hothouse*.

*ideology*: the body of beliefs, symbols, and sociopolitical aims of an individual, group, or institution.

*iğnorance*: the level of consciousness of unknowing, where nattermind runs living and inspires suffering for the resident consciousness. Compare *enlightenment*.

*immanent justice*: the belief in a natural force that enforces a moral universe. The concept of karma is exemplary.

*imagination*: the faculty for fabricating perceptions or symbolic constructs.

*Indus Valley Civilization* (aka *Harappan civilization*) (~9000–~3600 BF): a peaceful, prosperous civilization that flourished in the basins of the monsoon-fed Indus River, extending from northeast Afghanistan to northwest India. At its peak, the civilization may have had a population of over 5 million.

*Industrial Revolution*: the era of industrialization that began in England in the mid-4th century BF. English economic historian Arnold Toynbee popularized the term. Industrialization engendered 3 complementary social dynamics: 1) rapid urbanization, 2) a population boom, and 3) the destruction of the existing feudal social hierarchy, which was gradually replaced by a dominant social class of wealth inherited or made from manufacture, trade, and/or finance.

*intelligence* (biology): an attribution for behaving appropriately.

*interspecific*: occurring between distinct species. Contrast *conspecific*.

*Iron Age* (~3400–2600 BF): the last (3rd) principal period of the *3-age system*, noted for widespread use of iron and the development of steel. See *Stone Age*, *Bronze Age*.

## K

*karma*: the moralistic idea that Nature naturally rewards good and punishes evil in its own time (immanent justice).

*keystone species*: a species in a biome that acts as a provider or facilitator for other species, even indirectly.

*king*: a male monarch (sovereign).

## L

*labor theory of value*: the idea that the economic value of a good entirely derives from the labor involved to make it.

*language*: a system of symbols with interrelated meanings.

*legalism*: autocratic rule of law.

*Levant*: the broad geographical region east of the Mediterranean Sea. The term *Levant* first appeared in English in 603 BF, originally meaning "the East." The Levant was characterized as the "crossroads of western Asia, the eastern Mediterranean, and northeast Africa."

*life-history variable*: a trait or aspect of an organism's existence related to others; often viewed comparatively, as a trade-off with other, mutually exclusive possibilities.

*lithosphere*: the outermost shell of a rocky planet. Earth's lithosphere comprises its crust and upper mantle: the portions that behave elastically over geological expanses of time.

*localization* (physics): the process of creating a local effect or field from a universal field. Compare *quantization*.

## M

*mammal*: a class of air-breathing vertebrate animals, characterized by endothermy, hair, and females with mammary glands.

*manipularity-intelligence theory*: a theory by Ishi Nobu that the ease by which organisms can manipulate their environment is inversely related to acumen as a life-history variable.

*market economy*: a willy-nilly economic system operating via trade. Compare *command economy*, *traditional economy*.

*mass extinction*: the indiscriminate extinction of many species during a planetary extinction event.

*materialism* (psychology, economics): a worldview valuing material consumption and possessions.

*matter*: the mental fabrication of physicality. See *energy*.

*matterism*: the monistic belief that reality is made of matter. Matterism supposes that the mind is a figment of something substantial. Contrast *energyism*.

*meat*: animal tissue, especially the flesh of mammals, as food.

*medieval*: see *Middle Ages*.

*memory*: the recall of mental fabrications which are attributed as having occurred in the past.

*mentalizing* (aka *mind perception*): inferring the mental state of another being, typically another person.

*menopause*: the transitional period of reproductive senescence well before natural death, typically occurring in women around 50 years of age.

*Mesolithic* (20–5 TBF: 20–8 TBF in southwest Asia, 15–5 TBF in Europe): the final period of foraging in human cultures as the predominant lifestyle, before giving way to agriculture.

*Mesozoic–Cenozoic radiation* (beginning 65 MBF): a period of rapid radiation in new bird and mammal species. The fragmentation of the supercontinent Pangea was instrumental.

*methane* ($CH_4$) (aka *natural gas*): a colorless, odorless, flammable gas. Compare *coal*, *petroleum*.

*Middle Ages* (aka *medieval period*) (~1633–700 BF): the era of European history between the 17th and 7th centuries BF, beginning with the collapse of the (western) Roman Empire and the onset of the Dark Ages.

*mind*: an object-oriented misnomer for the mentation process.

*mind perception*: see *mentalizing*.

*Miocene* (23–5.3 MBF): the 1st of 2 epochs in the Neogene period, divided into 6 ages with corresponding geological stages.

*model* (mathematics): a mathematical construct. See *physical model*.

*monarch*: a commander, often an autocrat, in a dynasty.

*Mongols*: a central-east Asian ethnic group (race) first written about in the 14th century BF.

*Mongol conquests*: the incursion of Mongols into much of continental Asia during the 9th to 8th centuries BF.

*monism*: the metaphysical doctrine that there is a singular reality, either matterism or energyism. Contrast *dualism*.

*monkey*: a simian with a tail, as contrasted to tailless apes.

*monocot* (*monocotyledon*): a plant with a single embryonic leaf (cotyledon) in its seed. Other plant clades may have 2 (dicot) or 3 (eudicot) initial leaves.

*monocrop*: growing a single crop on a plot of land.

*monotheism*: belief in a singular god. Compare *polytheism*.

*moral*: conforming to a principle of appropriate behavior based upon respect for other life.

*moral absolutism*: the principle that acts are intrinsically right or wrong. See *deontology*.

*moral universe*: the belief that morality exists in Nature.

*multicellularity*: an organismal structure comprising multiple cells. Contrast *pluricellularity*.

## N

*naïve realism* (aka *commonsense realism*, *scientific realism*): the belief that actuality as perceived is reality.

*nation*: a political territory. Compare *state*. See *nation-state*.

*nation-state*: the government (state) of a nation; the concept of a nation and its governance as integral.

*nattermind* (aka *monkey-mind*): the involuntary, independent agent in the mind that instills and maintains iğnorance. Contrast *willmind*. See *coremind*.

*natural gas*: a methane-rich gas. Compare *coal*, *petroleum*.

*natural law*: a philosophy of law premised upon the belief that there a universal morality associated with fairness. See *moral universe*.

*natural philosophy*: the study of Nature from a holistic perspective; the common method of comprehending Nature until the 5th century BF, before science took over.

*naturalism*: the monistic matterist belief that observable actuality and reality are synonymous. Compare *naïve realism*. See *matterism*. Contrast *supernaturalism*.

*Nature*: the perceived exhibition of existence.

*Neogene* (23–0.05 MBF): the middle geological period of 3 in the Cenozoic era, during which mammals and birds evolved into their modern forms. Later in the period hominids arose.

*neoteny* (aka *juvenilization*, *pedomorphosis*): retention by adults of traits previously seen only in the young (in the perspective of evolutionary descent).

*New World*: the hemisphere of the Americas and nearby islands; sometimes Oceania is included. The term originated in the early 6th century BF by European explorers expanding their worldly horizons. Contrast *Old World*.

## O

*objectivity*: the idea that Nature and reality are independent of consciousness. Contrast *showtivity*.

*obsidian*: a naturally occurring volcanic glass formed when lava extruded from a volcano cools rapidly with minimal crystal growth.

*Oceania*: a region centered on the islands in the tropical Pacific Ocean, including Australasia.

*offshoring*: the relocation of a business process from one country to another.

*Old World*: Africa, Europe, and Asia: the part of the world known to Europeans prior to their sojourns to the Americas. Contrast *New World*.

*olfaction*: sense of smell.

*oligarchy*: despotic power by a privileged clique.

*Oligocene* (34–23 MBF): the 3rd and last geologic epoch in the Paleogene period.

*onager* (aka *Asiatic wild ass*): a donkey-like equid that evolved on the grasslands of Eurasia. Onagers were among the fastest mammals, able to run 65 km/h (40 mph).

*organic* (chemistry): a molecular structure based upon a carbon backbone.

*organism*: a cellular life form. Compare *virus*.

*Ottoman Empire* (801–178 BF): an empire founded by the Turks upon their conquest of Constantinople (647 BF), thus overthrowing the Byzantine Empire.

*oxygen (O)*: the element with atomic number 8; a highly reactive nonmetallic element that readily forms compounds (oxides) with almost all other elements.

## P

*Paleogene* (66–23 MBF): the 1st of 3 periods in the Cenozoic era.

*pandemic*: a widespread outbreak of a disease that afflicts many people on different continents.

*Pangea* (aka *Pangaea*): the supercontinent that contained most of Earth's land mass 300–200 MBF. The global ocean of the time was Panthalassa. Pangea broke up into Laurasia to the north and southern Gondwana.

*papal*: relating to the pope (ruler) of the Roman Catholic Church.

*parasitism*: the lifestyle of an organism living in, on, or with another organism in a way that typically affects the health of its host.

*parliament*: in modern times, a state lawmaking body of designated representatives which also often served as a check on the power of a ruler.

*peat*: an accumulation of partly decayed vegetation and other organic matter that becomes part of the soil. Areas rich in peat are called *peatlands*, *bogs*, *mires*, *moors*, or *muskegs*.

*perception*: the process of making sense of sensation, including determining significance and meaning.

*personal fact*: an individual construal. Compare *social fact*. See *fact*.

*pest*: an organism deemed a nuisance.

*pesticide*: a biocide intended to destroy pests.

*petroleum*: a natural yellow-to-black liquid comprising algae, zooplankton, and other organisms crushed, heated, and liquefied. Compare *coal*, *methane*.

*photosynthesis*: (an organism) converting sunlight into energy, using water and air as substrates.

*physical model*: a typically geometric or algebraic model yielding a mathematical characterization of an embodied phenomena.

*physical theory*: an explanation of relationships between various measurable phenomena. A physical theory may include a model of physical events (i.e., a physical model).

*phytonutrient*: a nutrient obtained by eating plants.

*plant*: a sessile autotroph. Mosses, ferns, conifers, and flowering plants (angiosperms) are plants. Compare *animal*.

*planthopper*: a tiny hopping insect that lived on vegetation, both residentially and dietarily. Planthoppers fed on the sugar-laden sap that courses through plant phloem.

*pluricellularity*: the structure of multiple cells aggregating in an organized manner. Contrast *multicellularity*.

*plutocracy*: (conservative) political rule by the wealthy.

*polar* (mathematics): a 2D coordinate system in which each point on a plane is referred to by a distance from a reference point and an angle from a reference direction. The reference point is the *pole*, and the ray from the pole the *polar axis*.

*polity*: a particular form of government.

*pollution*: something unhealthy lingering in the environment or causing a detriment in habitats.

*polytheism*: belief in multiple gods. Compare *monotheism*.

*precocial*: animals with relatively mature and mobile young from the moment of birth or hatching. Many, though not all, arthropods, fish, amphibians, and reptiles are precocial. Contrast *altricial*.

*primate*: a mammal order, containing prosimians (neither monkey nor ape) and simians (monkeys and apes).

*product*: an object made by labor. Compare *service*.

*prokaryote*: an organism that lacks a cell nucleus or other membrane-bound organelles. Bacteria are prokaryotes. While prokaryotes are single-celled, most can form stable, aggregate communities, such as a biofilm. Compare *eukaryote*.

*prophet*: someone with a cogent vision of the future.

## Q

*quantization* (physics): creating a localized appearance of matter. Compare *localization*.

*quorum sensing*: decision-making in a decentralized network.

## R

*radiation* (evolutionary biology): profuse adaptive speciation.

*reality*: that which necessarily is, regardless of existence.

*realization*: see *unity consciousness*.

*Roaring Twenties*: the (Gregorian) 1920s (170s BF) decade in Western culture, especially in the USA. A facile prosperity bloomed from financial speculation, which crashed at the end of the decade, ushering the Great Depression.

*rodent*: an order of mammals characterized by constantly growing incisors that must be kept short by gnawing.

*ruler*: a commander with limited power, which may be checked by a parliament or other authority in a state.

*ruminant*: a hoofed mammal herbivore that digests via stomach fermentation. Bovids are ruminants.

## S

*saltation*: a leap in evolution.

*sand puppy* (aka *naked mole rat, desert mole rat*): a eusocial burrowing rodent native to East Africa.

*savanna* (aka *savannah*): a grassland biome with trees sufficiently spaced so that the canopy does not close, despite a tree density that may be greater than a forest.

*science*: the study of Nature, emphasizing empirical methods. Contrast *natural philosophy*.

*scientific method*: a set of empirical techniques for investigating phenomena, ostensibly involving careful observation before guessing what is going on, which is known as forming a *theory*. Guessing prior to extensive and intensive observation is making a *hypothesis*.

*self-organized criticality*: a property of dynamic systems where a critical threshold (tipping point) exists that, when passed, sets off a substantial reaction.

*sensation, sense*: uptake from the senses. See *perception*.

*service*: an economic activity of benefit to someone.

*showtivity*: a seemingly shared experiential platform provided by Ĉocö as an ordering principle for the perception of Nature. Contrast *objectivity*.

*simian*: the suborder of primates comprising the "higher" primates: monkeys, apes, and humans. Simians tend to be larger than prosimians ("lower primates").

*simple supernaturalism*: the monistic doctrine that all of Nature is a unicity of entangled interaction. Compare *animism*.

*smelt*: applying heat to an ore to extract a base metal. Smelting was used to extract valued metals from the ores in which they were mined, including silver, iron, copper, and other base metals. The first metals smelted were lead and tin, beginning around 10.2 TBF. Smelting severely pollutes nearby water, soil, and air.

*smog*: lingering foul air; a portmanteau of *smoke* and *fog*.

*social fact*: a construal accepted by others. Compare *personal fact*. See *fact*.

*socialism*: an economy aimed at equity. To avert externalities, socialist theorists favored a command economy. Contrast *capitalism*.

*solipsism*: the irrefutable argument that only oneself can be proven to exist; universally ignored by Earthers.

*sovereign debt*: debt incurred by a state from spending more than (tax) revenues garnered; debt via deficit spending. Practicing unsustainable governance, modern states incurred stunning levels of sovereign debt.

*speciation*: the process of species formation.

*species* (biology): a distinct population of organisms.

*starling*: a (bird) family of sociable, medium-sized passerines. Many Asian starling species, particularly larger ones, were called *mynas*.

*state* (politics): the governing group (clique) of a nation.

*status function*: a mental attachment to an object by ascribing it esteemed meaning, purpose, or function.

*steppe*: a biome of grassland plains, lacking trees except near rivers and lakes.

*Stone Age* (roughly 3.4 MBF–5400 BF): the 1st principal period of the 3-age system, noted for use of stone tools, prior to the advent of metalworking. See *Bronze Age*, *Iron Age*.

*subjective*: the perspective of an individual mind.

*sulfur* (*S*): abundant on Earth, sulfur is a key element in biological processes. As an atmospheric pollutant (combined with oxygen), sulfur is a source of toxic acid rain.

*Sun*: the star at the center of the planetary system to which Earth belonged. The Sun has a diameter of 1,392,000 km.

*supernaturalism*: the belief that energetic forces are in play beyond the phenomenal. See *simple supernaturalism*, *energyism*. Contrast *naturalism*.

*sustainability*: the fanciful Earther concept of controllably exploiting Nature so that innumerable future generations of humans could prosper.

## T

*tactility*: sense of touch.

*talionic*: based upon retribution.

*tectonic plate*: a sizable slab of the lithosphere, including some of Earth's crust, capable of movement.

*teleology* (evolutionary biology): the amply evidenced theory that adaptation is goal oriented.

*teleology* (moral philosophy) (aka *consequentialism*): adjudging morality by the outcome of acts. Contrast *deontology*.

*teleology* (philosophy): the doctrine that final causes (ends or purposes) exist.

*theory*: fact-based explanation about the relations between concepts. Compare *hypothesis*. See *physical theory*.

*tipping point*: a critical moment in a complex situation in which events produce an inevitable change in the vector of future events.

*traditional economy*: a survival system with scant trade. Compare *command economy*, *market economy*.

*travois*: a transport device comprising 2 poles serving as shafts, joined to a frame, and bearing a platform for the load.

*tribe*: a social group with a shared culture. Compare *clan*.

*turbine*: a machine capturing energy from a moving fluid for work.

*Turk*: a member of numerous Asian peoples speaking a Turkic language.

*turtle*: a reptile with a unique, defensive, bony shell developed from its ribs, in the Testudines order. Originating 220 MBF via saltation, turtles were one of the oldest reptile groups.

## U

*ungulate*: a group of mammals which use the tips of their toes, typically hoofed, to sustain body weight while moving. Even-toed ungulates (artiodactyl) bear their weight equally between the 3rd and 4th toes. Odd-toed ungulates (perissodactyl), which have an odd number of toes on their rear feet, bear weight on their 3rd toe.

*unity consciousness* (aka *realization*): the supreme level of consciousness, with an abiding experience of the unicity of Nature via connectivity with Cönsciousness. Compare *enlightenment*, see *iğnorance*.

## V

*virus*: an obligate parasite that self-evolved to infect cells in all other life. Viruses aided their venture by inventing DNA and introducing it to organisms.

*vision*: the sense of sight through light.

*vitalism* (biology): the doctrine that there is a vital energy to life, distinct from chemical and physical forces.

*vitalism* (natural philosophy): the idea that life is essentially distinct from inanimate matter. Compare *animism*.

## W

*weather*: characterization of daily or other short-term tropospheric conditions in a locality. Compare *climate*.

*Western* (*world*) (politics): nations in Europe and the Americas. Derived from the Latin *occidentalis*, which meant *western*. In the 7th century BF, Europeans contrasted the Occident (the West) from the Orient (the East). Whereas the West comprised Europe and its conquests in the Americas, the East included China, Japan, and India.

*white* (sociology): modern vernacular for light-skinned people, especially those with a European heritage. Contrast *black*.

*willmind*: willful (volitional) mentation. Contrast *nattermind*.

## Z

*zoonosis*: an infectious, pathogenic (animal) disease that can be transmitted to humans. The major modern human diseases were zoonoses. Zoonoses may be sourced to a wide range of pathogens, such as viruses, bacteria, fungi, and parasites. Most pathogenic human diseases originated in other animals. But only diseases that were routinely transmitted from other animals to humans were considered zoonosis. When humans infect other animals, the term used was *anthroponosis* or *reverse zoonosis*.

# Acknowledgments

Earth satellite images adapted from photos by NASA via eChalk (eChalk.co.uk).

Planthopper photos courtesy of Malcolm Burrows, zoologist at the University of Cambridge, UK.

Political maps courtesy of MapChart (mapchart.net).

Electricity provision map adapted from a NASA satellite photo.

World shipping routes courtesy of Benjamin S. Halpern, Tomislav Hengl, and Grolltech.

Coral reefs map derived from NASA data.

Last and foremost, this work was made possible by the generosity of Sarah Jean Wolf.

# Index